Get Inspired!

☑ CHECK IT OFF
Make sure to see these inspiring features as you review this program guide!

☐ A Next Generation Instructional Model
Take a close look at the Module and Lesson Design on **pages 10–11** to see how *Inspire Science* is designed for three-dimensional learning.

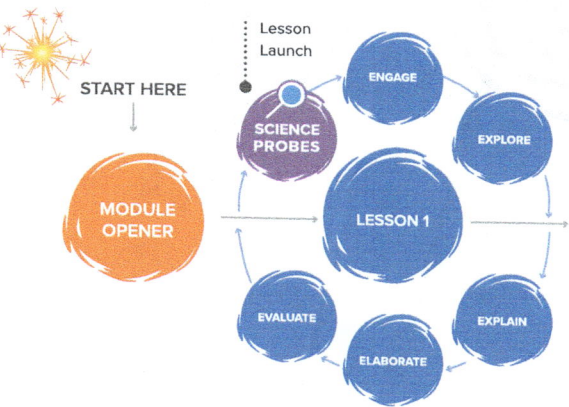

☐ Rethinking Opportunities
With *Inspire Science*, your students will think, investigate, and rethink in every lesson—just like real-world scientists and engineers do. Look for these examples of these circling back opportunities on these pages:

- Collect Evidence Prompts and the CER Framework, **pages 47** and **50**
- Revisit the Science Probe, **page 43**
- Explain the Phenomenon, **page 52**

Each *Inspire Science* lesson begins with a Formative Assessment Science Probe.

☐ Phenomena-Driven Learning
See how phenomena drive the *Inspire Science* learning experience on **page 14.**

ENCOUNTER THE PHENOMENON
How do the goats climb the tree?

☐ Research-Driven Inquiry Approach
Take a look at **page 16** to learn about the advanced and research-based approach to inquiry-based learning that's at the center of the student-led learning experience in *Inspire Science*.

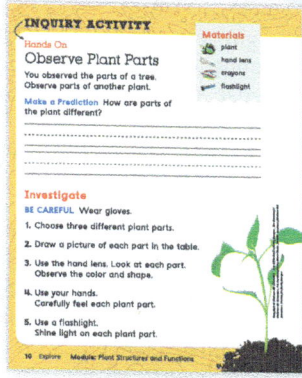

☐ Next Generation Assessments
As you'll see on **page 26**, with *Inspire Science* you can be confident that you have a program that guides students down a path to success with the Performance Expectations.

Measured Progress is the leading provider of assessments designed specifically for NGSS. These test items are in every *Inspire Science* module.

Get Inspired! 1

my.mheducation.com

Copyright © 2020 McGraw-Hill Education

All rights reserved. No part of this publication may be reproduced or distributed in any form or by any means, or stored in a database or retrieval system, without the prior written consent of McGraw-Hill Education, including, but not limited to, network storage or transmission, or broadcast for distance learning.

Permission is granted to reproduce the material contained in this book on the condition that such material be reproduced for classroom use only; be provided to students, teachers, or families without charge; and be used solely in conjunction with *Inspire Science*.

Send all inquiries to:
McGraw-Hill Education
8787 Orion Place
Columbus, OH 43240

Program Guide:
ISBN: 0-07-700400-0
MHID: 978-0-07-700400-2

Printed in the United States of America.

9 10 11 12 MER 27 26 25 24

COV (t)Dazzo/Radius Images/Getty Images, (cl)Butterfly Hunter/Shutterstock, (cr)Hero Images/Getty Images, (b)FreedomMaster/iStock/Getty Images

Our mission is to provide educational resources that enable students to become the problem solvers of the 21st century and inspire them to explore careers within Science, Technology, Engineering, and Mathematics (STEM) related fields.

Program Design

Welcome

Explore Our Phenomenal World

Curiosity drives learning. *Inspire Science* provides an in-depth, collaborative, and project-based learning experience designed to help you spark students' interest and empower them to ask more questions and think more critically. Through inquiry-based, hands-on investigations of phenomena, your students will answer more rigorous science questions with evidence and generate innovative solutions to real-world problems.

Are you ready to inspire the next generation of innovators?

 **Inspire Curiosity**
Spark critical thinking.

100%
Built for the Next Generation Science Standards (NGSS)

 **Inspire Investigation**
Spark inquiry-driven, hands-on exploration.

Inspire Innovation
Spark creative solutions to real-world challenges.

 Need login credentials?
Go to my.mheducation.com and select "Create Teacher Account."

Key Shifts for NGSS Success

Next Generation Science Standards are designed to help prepare students for career and college readiness through a more innovative approach to K-12 science education. This new approach requires a few shifts in science instruction and learning, and *Inspire Science* supports you through each one.

 Look for this symbol throughout this guide to learn more about these *Key Shifts for NGSS Success*:

- Three-Dimensional Learning
- Depth Over Breadth
- Phenomena-Driven, Inquiry-Based, Hands-On Learning
- Evaluating Performance Over Testing Knowledge
- Integrated Engineering
- Progressive Learning

 ## Three-Dimensional Learning

The three-dimensional learning framework of *Inspire Science* delivers on the application-oriented approach needed to prepare your students for any challenge.

SEP Science and Engineering Practices

SKILLS
(for example, "Developing and Using Models")

DCI Disciplinary Core Ideas

CONTENT IN FOCUS
(for example, "The Universe and Its Stars")

CCC Crosscutting Concepts

COMMON THEMES
(for example, "Systems and System Models")

Inspire Science

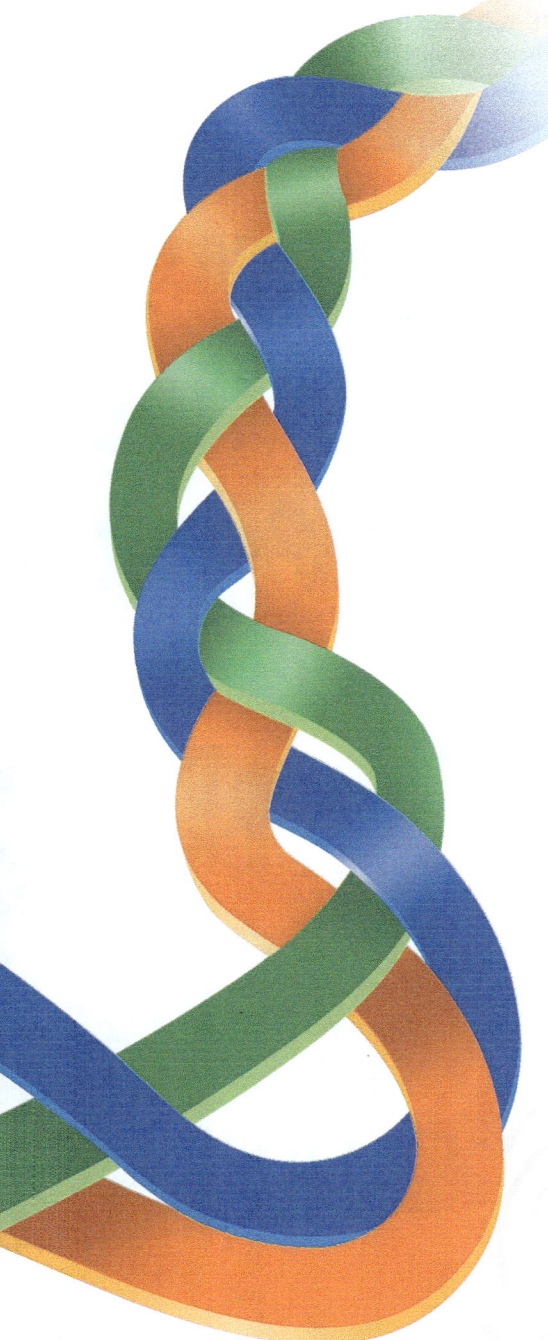

 ### Performance Expectations

These statements describe what students must actually do in order to demonstrate mastery of a subject area's core content.

Students achieve proficiency with the Performance Expectations by working with the Science and Engineering Practices, Disciplinary Core Ideas, and Crosscutting Concepts in tandem to make sense of phenomena and design solutions to real-world problems.

(for example, "Use observations of the Sun, Moon, and stars to describe patterns that can be predicted.")

CROSS-CURRICULAR Connections

The *Inspire Science* lessons include cross-curricular connections with quick and easy references to the specific literacy and math skills being reinforced through the science investigations.

Science and Engineering Handbook

 **Go Online** Use the digital Science and Engineering Handbook to learn more about each of the eight science and engineering practices and crosscutting concepts, as well as helpful science and engineering background information.

Key Shifts for NGSS Success

Depth Over Breadth

Inspire Science students will shift from a wide range of topics with shallow exploration to a more narrow range of topics with in-depth exploration to advance conceptual understanding.

TRADITIONAL APPROACH
Wide Range of Topics, Shallow Exploration

THE NGSS APPROACH
More Narrow Range of Topics, In-Depth Exploration

Phenomena-Driven, Inquiry-Based, Hands-On Learning

Students build long-lasting knowledge and skills by experiencing science and engineering in a more meaningful, real-world, application-oriented way. *Inspire Science* delivers on this approach through:

- Phenomena-Driven Learning
- Inquiry-Based Learning
- Hands-On Learning
- Project-Based Learning

DISCOVER
THE PHENOMENON

> What happens when you blow on a dandelion?

Inspire Science

Evaluating Performance Over Testing Knowledge

The formative and summative assessments in *Inspire Science* focus on helping students achieve a deep level of conceptual understanding performance-based evaluations and rubrics.

Integrated Engineering

One of the key shifts in the NGSS is the addition of the engineering design strand. Engineering activities and content (and teacher support) are seamlessly integrated throughout *Inspire Science*.

Progressive Learning

The NGSS progressions build on concepts year after year to deepen conceptual understanding over time. These progressions serve as a key building block for *Inspire Science*, allowing students to learn more about a given topic each year for an in-depth understanding by the end of Grade 12.

K-2
Patterns of the motion of the sun, moon, and stars in the sky can be observed, described, and predicted.

3-5
The sun is a star that appears larger and brighter than other stars because it is closer. Stars range greatly in their distance from Earth.

6-8
Earth and its solar system are part of the Milky Way galaxy, which is one of many galaxies in the universe.

9-12
The star called the sun is changing and will burn out over a lifespan of approximately 10 billion years.

College and Career Ready!

Disciplinary Core Idea Progression: The Universe and Its Stars

Key Shifts for NGSS Success

Scope and Sequence, K-5 Integrated

Grade K

UNIT 1	**LIVING THINGS**
MODULE	Plants and Animals
LESSON 1	Living and Nonliving
LESSON 2	Plant and Animal Survival
LESSON 3	Places Plants Live
LESSON 4	Places Animals Live
UNIT 2	**OUR CHANGING WORLD**
MODULE	Changes to the Environment
LESSON 1	Plants Change Their Environment
LESSON 2	Animals Change Their Environment
LESSON 3	People Change Their Environment
MODULE	Protect Earth
LESSON 1	Natural Resources
LESSON 2	Reduce, Reuse, Recycle
UNIT 3	**WEATHER AND THE SUN**
MODULE	Weather
LESSON 1	Describe Weather
LESSON 2	Weather Patterns
LESSON 3	Forecast Weather
LESSON 4	Severe Weather
MODULE	The Sun and Earth's Surface
LESSON 1	Sunlight on Earth's Surface
LESSON 2	Protection from the Sun
UNIT 4	**MAKE THINGS MOVE**
MODULE	Forces and Motion
LESSON 1	Pushes and Pulls
LESSON 2	Direction and Speed
LESSON 3	When Objects Collide

Grade 1

UNIT 1	**ALL ABOUT PLANTS**
MODULE	Plant Structures and Functions
LESSON 1	Plant Parts
LESSON 2	Functions of Plant Parts
MODULE	Plant Parents and Their Offspring
LESSON 1	Plants and Their Parents
LESSON 2	Plant Survival
UNIT 2	**ANIMALS AND HOW THEY COMMUNICATE**
MODULE	Animals Parents and Their Offspring
LESSON 1	Animal Structures
LESSON 2	Functions of Animal Structures
LESSON 3	Animals and Their Parents
LESSON 4	Animal Behaviors
MODULE	Communication
LESSON 1	Animal Communication
LESSON 2	Sound
UNIT 3	**LIGHT AND SHADOWS**
MODULE	See Objects
LESSON 1	Light
LESSON 2	Light and Materials
LESSON 3	Light Uses
UNIT 4	**SKY PATTERNS**
MODULE	Observe the Sky
LESSON 1	Objects in the Sky
LESSON 2	Day and Night Patterns
LESSON 3	Patterns During the Year

Grade 2

UNIT 1	**LAND AND WATER**
MODULE	Earth's Landscape
LESSON 1	Local Landscapes
LESSON 2	Land and Earth
LESSON 3	Water on Earth
UNIT 2	**PROPERTIES OF MATERIALS**
MODULE	Describe Materials
LESSON 1	Investigate Materials
LESSON 2	Test and Analyze Materials
MODULE	Changes to Materials
LESSON 1	Build with Materials
LESSON 2	Materials Can Change
UNIT 3	**EARTH'S CHANGING LANDSCAPE**
MODULE	Landscape Changes
LESSON 1	Slow Changes to Earth's Landscape
LESSON 2	Quick Changes to Earth's Landscape
LESSON 3	Design Solutions to Slow Landscape Changes
UNIT 4	**LIVING THINGS AND HABITATS**
MODULE	Plants in Landscapes
LESSON 1	What Plants Need
LESSON 2	Plants Depend on Animals
MODULE	Living Things in Habitats
LESSON 1	Local Habitats
LESSON 2	Land Habitats
LESSON 3	Water Habitats

K-5 Learning Progression within Each Grade

Inspire Science modules are bundled in a sequence designed to support learning progression toward the grade-level Performance Expectations in alignment with the NGSS. The progressions within each grade establish a strong base of knowledge for the Performance Expectations the following years.

Inspire Science

Grade 3

UNIT 1 — FORCES AROUND US

MODULE: Forces and Motion
- LESSON 1: Motion
- LESSON 2: Forces Can Change Motion

MODULE: Electricity and Magnetism
- LESSON 1: Electricity and Designing Solutions
- LESSON 2: Magnetism and Designing Solutions

UNIT 2 — LIFE CYCLES AND TRAITS

MODULE: Plants
- LESSON 1: Plant Life Cycles
- LESSON 2: Plant Traits

MODULE: Animals
- LESSON 1: Animal Life Cycles
- LESSON 2: Animal Traits
- LESSON 3: Animal Group Survival

UNIT 3 — DIFFERENT ENVIRONMENTS

MODULE: Survive the Environment
- LESSON 1: Survival of Organisms
- LESSON 2: Adaptations and Variations

MODULE: Change the Environment
- LESSON 1: Fossils
- LESSON 2: Changes Affect Organisms

UNIT 4 — OBSERVING WEATHER

MODULE: Weather Impacts
- LESSON 1: Weather Patterns
- LESSON 2: Weather and Seasons
- LESSON 3: Natural Hazards and the Environment
- LESSON 4: Prepare for Natural Hazards

Grade 4

UNIT 1 — FORCES AND ENERGY

MODULE: Energy and Motion
- LESSON 1: Forces and Motion
- LESSON 2: Speed and Energy
- LESSON 3: Energy Transfer in Collisions

UNIT 2 — USING ENERGY

MODULE: Energy Transfer
- LESSON 1: Types of Energy
- LESSON 2: Sound and Light
- LESSON 3: Electricity
- LESSON 4: Heat

MODULE: Natural Resources in the Environment
- LESSON 1: Energy from Nonrenewable Resources
- LESSON 2: Energy from Renewable Resources
- LESSON 3: Impact of Energy Use
- LESSON 4: Design Energy Solutions

UNIT 3 — OUR DYNAMIC EARTH

MODULE: Earth and Its Changing Features
- LESSON 1: Map Earth's Features
- LESSON 2: Evidence from Rocks and Fossils
- LESSON 3: Changes in Landscapes Over Time

MODULE: Earthquakes
- LESSON 1: Map Earthquakes
- LESSON 2: Model Earthquake Movement
- LESSON 3: Reduce Earthquake Damage

UNIT 4 — INFORMATION PROCESSING AND LIVING THINGS

MODULE: Structures and Functions of Living Things
- LESSON 1: Structures and Functions of Plants
- LESSON 2: Structures and Functions of Animals

MODULE: Information Processing and Transfer
- LESSON 1: Information Processing in Animals
- LESSON 2: Role of Animals' Eyes
- LESSON 3: Information Transfer

Grade 5

UNIT 1 — INVESTIGATE MATTER

MODULE: Matter
- LESSON 1: Identify Properties of Materials
- LESSON 2: Mixtures and Solutions
- LESSON 3: Physical and Chemical Changes
- LESSON 4: Solids, Liquids, and Gases

UNIT 2 — ECOSYSTEMS

MODULE: Matter in Ecosystems
- LESSON 1: Plant Survival
- LESSON 2: Interactions of Living Things
- LESSON 3: Role of Decomposers

MODULE: Energy in Ecosystems
- LESSON 1: Earth's Major Systems
- LESSON 2: Cycles of Matter in Ecosystems
- LESSON 3: Energy Transfer in Ecosystems

UNIT 3 — EARTH'S INTERACTIVE SYSTEMS

MODULE: Earth's Water System
- LESSON 1: Water Distribution on Earth
- LESSON 2: Human Impact on Water Resources
- LESSON 3: Effects of the Hydrosphere

MODULE: Earth's Other Systems
- LESSON 1: Effects of the Geosphere
- LESSON 2: Effects of the Atmosphere
- LESSON 3: Effects of the Biosphere

UNIT 4 — EARTH AND SPACE PATTERNS

MODULE: Earth's Patterns and Movement
- LESSON 1: The Role of Gravity
- LESSON 2: Earth's Motion

MODULE: Earth and Space
- LESSON 1: Earth's Place in Space
- LESSON 2: Stars and Their Patterns

Module Experience At A Glance

Inspire Science's phenomena-driven 5E lessons are designed to provoke critical thinking and spark creative problem solving.

Grades K–1 Module and Lesson Design

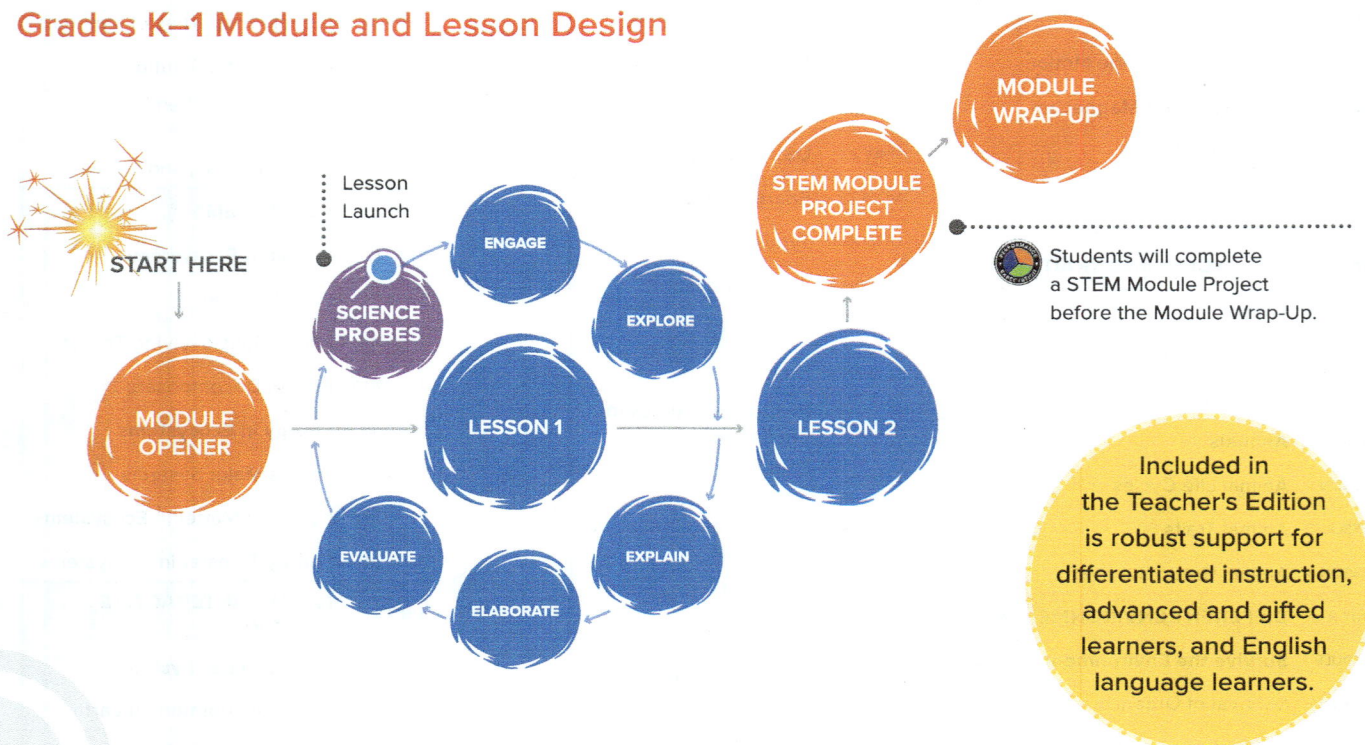

Pacing Options to Fit Your Schedule

FullTrack
45 min/day (5 days a week)

FlexTrack A
30 min/day (5 days a week)

FlexTrack B
30 min/day (3 days a week)

Key Student Activities

 STEM Module Project Launch (Grades 2–5)

MODULE OPENER	ASSESS PRIOR KNOWLEDGE	ENGAGE	EXPLORE
Discover / Encounter the Phenomenon	Science Probe Formative Assessment	Discover / Encounter the Phenomenon	Explore the Phenomenon (Inquiry Activity)
STEM Connection		Talk About It	Claim, Evidence, Reasoning (CER) (Grades 2–5)
Talk About It			Cross-Curricular Connections
Word Wall (Grades K–1)			
Module Pretests (G2–5)			

10 Program Design

Inspire Science

Grades 2–5 Module and Lesson Design

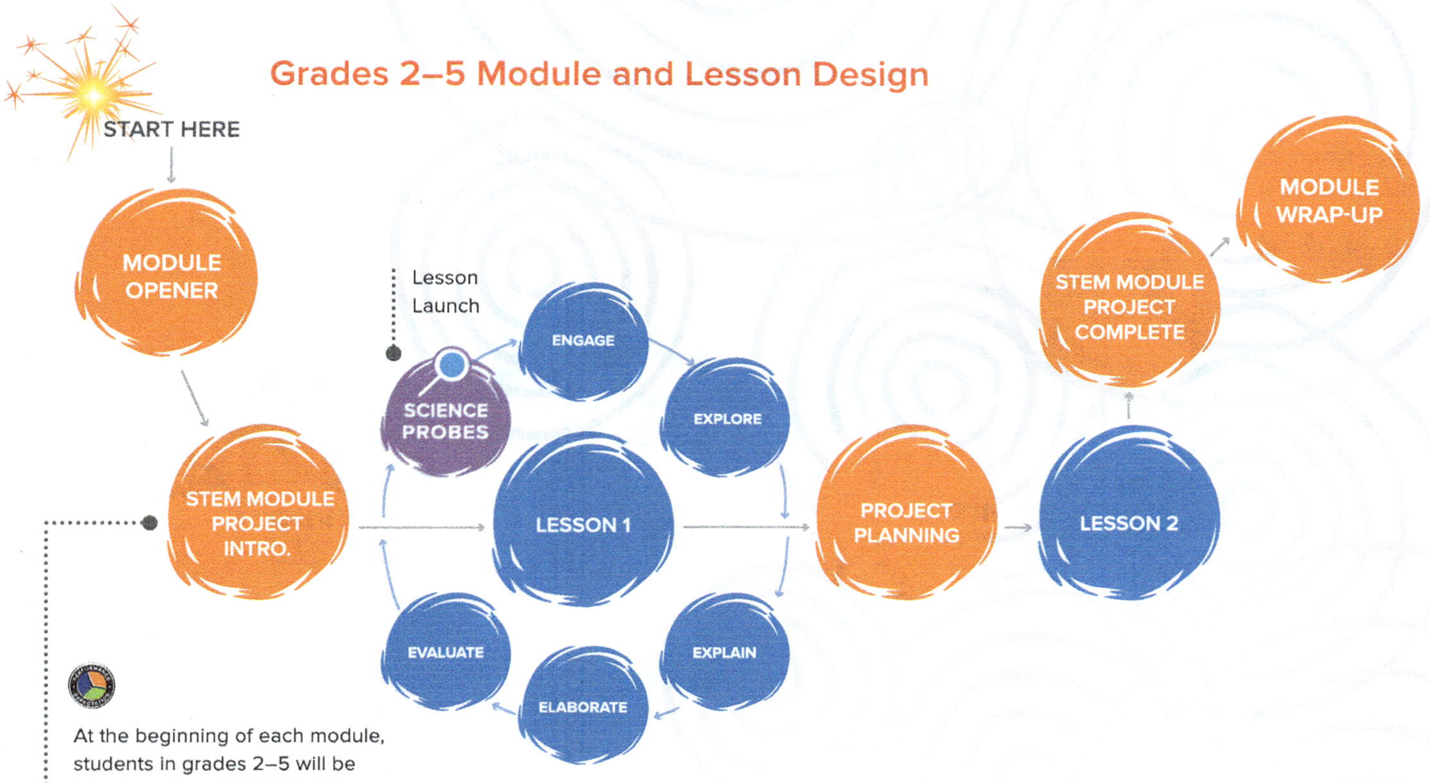

START HERE → MODULE OPENER → STEM MODULE PROJECT INTRO.

Lesson Launch → SCIENCE PROBES

LESSON 1: ENGAGE → EXPLORE → EXPLAIN → ELABORATE → EVALUATE

→ PROJECT PLANNING → LESSON 2 → STEM MODULE PROJECT COMPLETE → MODULE WRAP-UP

At the beginning of each module, students in grades 2–5 will be introduced to a STEM Module Project that they will complete at the end of the module. Touch points at the end of each lesson provide for project planning.

STEM Module Project Planning (after each lesson in Grades 2–5) and Completion (end of the module in Grades K–5)

EXPLAIN	ELABORATE	EVALUATE	MODULE WRAP-UP
Vocabulary	Inquiry Activities	Lesson Review	Rediscover / Revisit the Module Phenomenon
Inquiry Activities	STEM Connection	Explain the Phenomenon	Three-Dimensional Assessment
Close Reading	Environmental Connections	Revisit the Science Probes	
Talk About It	Close Reading	Three-Dimensional Assessment	
Revisit the Science Probe	Three-Dimensional Questions		
Three-Dimensional Thinking			
Claim, Evidence, Reasoning (CER) (Grade 1)			
Cross-Curricular Connections			
Quick Check			

Module Experience At-A-Glance 11

Resources At-A-Glance

Print Resources

Each interactive Student Edition unit encourages hands-on learning through the NGSS and Framework. Each Teacher Edition unit provides in-depth teacher strategies to make sure your classroom succeeds.

TEACHER'S EDITION
Available in Spanish (Grades K–5, Four Units Per Grade)

Unit 4, Unit 3, Unit 2, Unit 1

STUDENT EDITION
Available in Spanish (Grades K–5, Four Units Per Grade)

Unit 4, Unit 3, Unit 2, Unit 1

SCIENCE READ ALOUDS
Available in Spanish (Grades K–1)

INVESTIGATOR ARTICLES
Available in Spanish (Grades 2–5)

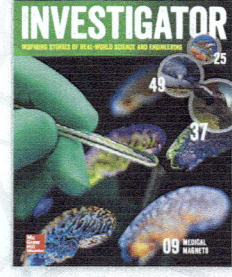

- Approaching Level (online, printable)
- On Level

LEVELED READERS
Available in Spanish (Grades K–5)

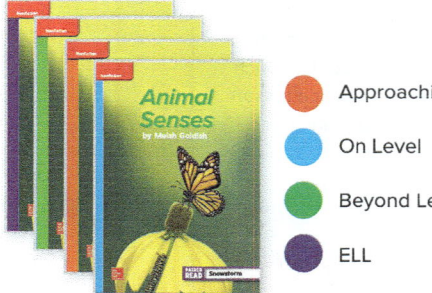

- Approaching Level
- On Level
- Beyond Level
- ELL

Collaboration Kits

(for small group Hands-On Inquiry Activities)

Inspire Science Collaboration Kits make planning for hands-on time easier so you can focus on the activities. Each Collaboration Kit contains the materials needed for the hands-on inquiry activities, organized by unit and module.

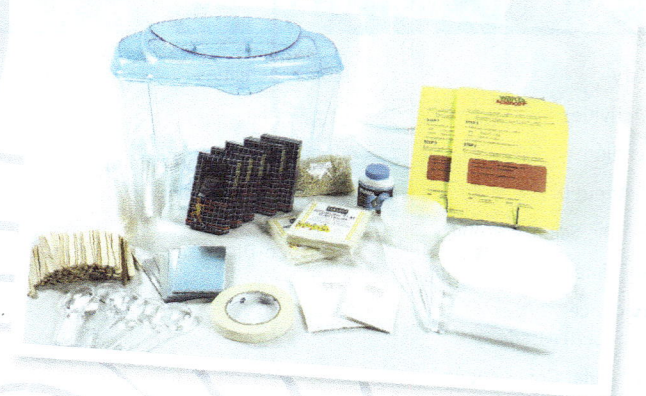

Inspire Science

Student Digital Resources

🖱 Why Go Online?

- Engaging Interactive Content
- Video Demos of Hands-On Activities
- Science Content Videos
- Text Read Aloud and Highlighting Features
- Dynamic Search Tools
- Impact News

Print books include digital versions with interactive features, including audio and text highlighting.

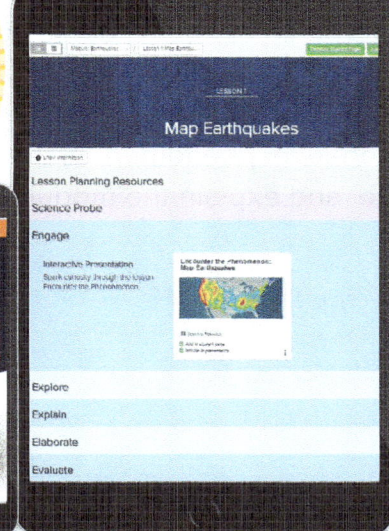

Spanish Digital Center Also Avaible

Type Entry
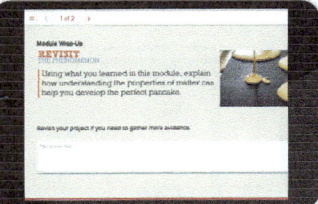

Drawing Tool

Drag and Drop
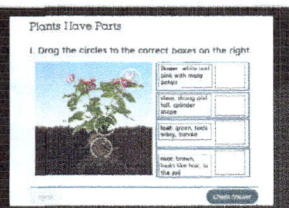

Simulations

Phenomena Videos

Science Content Videos

Personal Tutors (4-5)
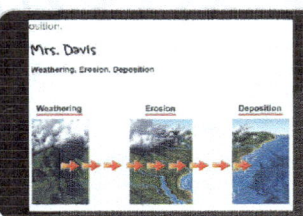

Impact News

Beyond the Classroom (2–5)

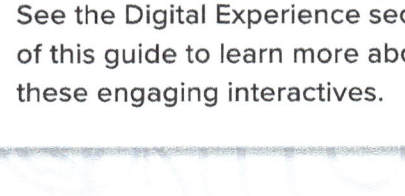

See the Digital Experience section of this guide to learn more about these engaging interactives.

Resources At-A-Glance 13

Phenomena-Driven Learning

Every day, we are surrounded by natural phenomena that pique our curiosity. In *Inspire Science*, these phenomena are the centerpiece of each module and lesson to engage students and inspire them to investigate key science and engineering concepts through their three-dimensional learning experience. As students investigate each lesson-level phenomenon, they will gather pieces of the puzzle to help solve and explain the module-level phenomenon.

ENCOUNTER THE PHENOMENON

How do the goats climb the tree?

Anchoring Module Phenomena

Investigative Lesson Phenomena

Students will investigate related lesson-level phenomena that will help them build understanding so they can uncover the question of the anchoring module phenomena.

LESSON 1

Will the cub grow up to look more like the adult mountain lion?

LESSON 2

Why do the kittens look different from the mom and each other?

LESSON 3

Why are the fish swimming in a circle?

Revisit the Phenomenon

In the Module Wrap-Up, students will connect what they've learned through the investigative lesson phenomena to explain the anchoring module phenomenon.

Phenomena-Driven, Inquiry-Based, Hands-On Learning

Inquiry-Based Learning

An inquiry-based approach to science and engineering education helps spark student curiosity and empower them to ask more questions, think more critically, answer deeper questions, and design solutions to the problems in their world. Today's students will need to know how to investigate questions and solve problems from a variety of angles. Inquiry-driven instruction gives students the practice they need to succeed in developing solutions to whatever challenges they may encounter.

In *Inspire Science*, students will conduct two to three inquiry activities per lesson, typically in the Explore, Explain, and Elaborate phases of the 5E model. Students will use their results and findings from each lesson to communicate their understanding through the STEM Module Project at the end of each module. These activities help students achieve proficiency with the science and engineering practices disciplinary core ideas, and crosscutting concepts.

Types of Inquiry Activities in *Inspire Science*

Inquiry is more than hands-on activities. With *Inspire Science*, students will investigate phenomena using the same techniques and practices that scientists and engineers use.

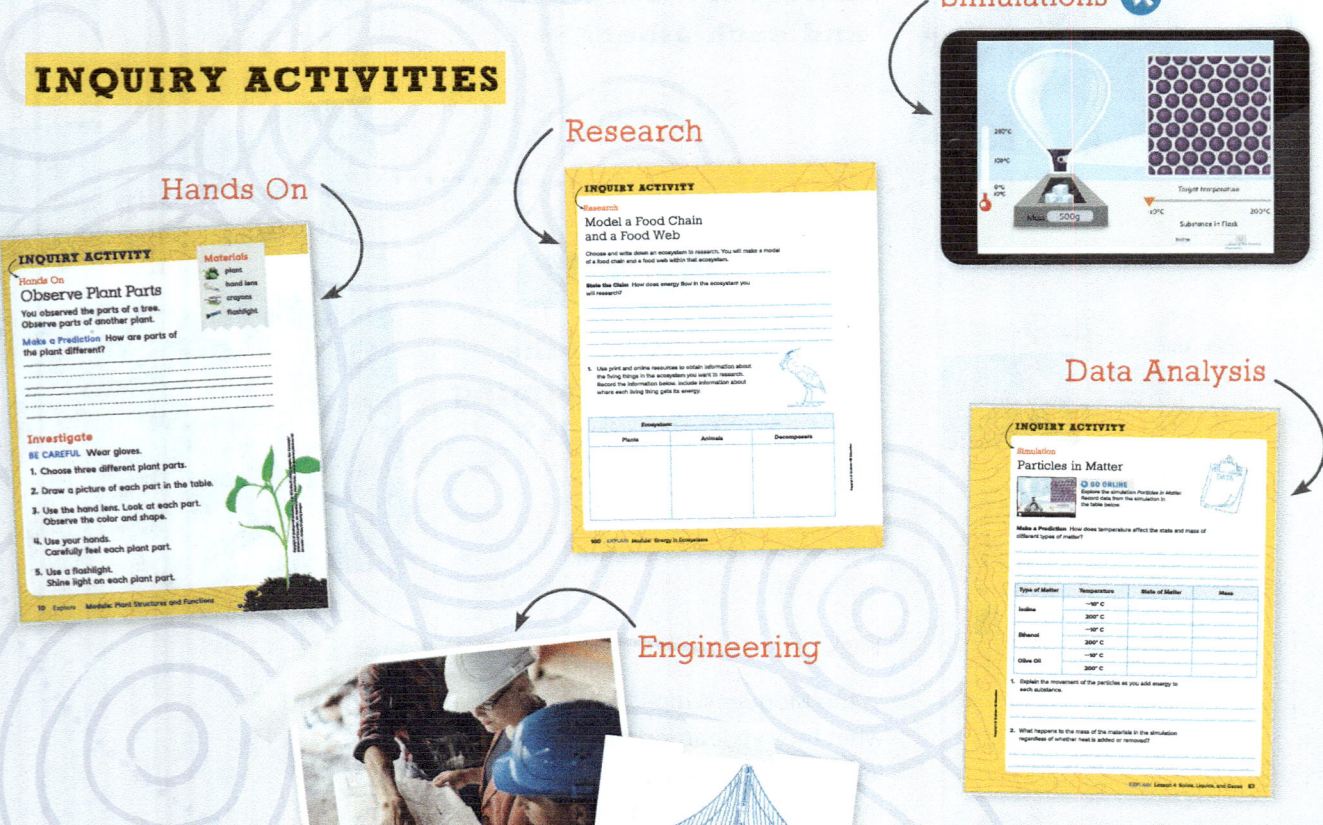

INQUIRY ACTIVITIES

- Hands On
- Research
- Simulations
- Engineering
- Data Analysis

Inspire Science

ENCOUNTER
THE PHENOMENON

Why don't mastodons exist anymore?

The *Inspire Science* Inquiry Spectrum

Not all inquiry activities are the same. Depending upon the available time and student readiness, structured inquiry might be perfect or your class may be ready for open inquiry. The *Inspire Science* **Inquiry Spectrum** provides flexible options to adjust the inquiry level to align with the learning needs of each student.

Each lesson offers inquiry activities that have been developed with a recommended inquiry spectrum level, giving you the flexibility to modify the level of instruction based on your students' needs. The Inquiry Spectrums are provided in the Teacher's Edition and online at point of use in the teacher support for the lesson.

Inquiry Spectrum

Structured Inquiry
This activity it **Structured Inquiry.**

Guided Inquiry
Provide students with the explorable question and the prediction. Have students write their own procedure.

Open Inquiry
Remind students of the phenomenon, and allow time for students to continue their research on ramps. Bookmark appropriate websites and provide quality texts for students to continue their investigations.

Phenomena-Driven, Inquiry-Based, Hands-On Learning 17

Hands-On Learning

Inspire Science uses hands-on inquiry activities designed to engage students, inspire investigation, and motivate deeper thinking about core science concepts—without creating a logistical burden for you. To make hands-on time a little easier, *Inspire Science* includes:

- Neatly organized **Collaboration Kits** with hands-on materials
- Inquiry Activity Support Videos
- **Module Inquiry Activity Planners** in the Teacher's Edition

Collaboration Kits with Customer Support

Inspire Science Collaboration Kits make planning for hands-on time easier so you can focus on the activities. Each Collaboration Kit contains most of the materials needed for the hands-on inquiry activities, organized by unit and module. Materials are clearly labeled and correlated with each lesson.

Inquiry Activity Support Videos

Every EXPLORE Inquiry Activity is paired with an Inquiry Rewind video to encourage student engagement. In these videos, students are provided guidance through every step of every activity. These videos are perfect for students who might have missed the in-class activity or who might be struggling to achieve the expected outcomes.

Every Inquiry Rewind video:

- Shows the activity materials and the step-by-step procedure
- Demonstrates the expected observations for each step of the activity
- Provides opportunities for students to pause the video and utilize their Claim, Evidence, and Reasoning skills they have learned through *Inspire Science*.

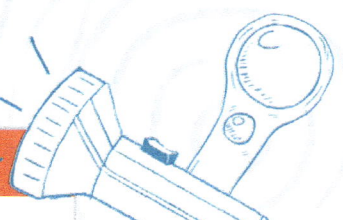

Inquiry Activity Planners

Planning and preparing for inquiry activities is made easier with the *Inspire Science* Inquiry Activity Planners. The planners clearly identify all the hands-on materials needed throughout the module and which materials are found in your *Inspire Science* Collaboration Kits.

Inquiry Activity Support

GO ONLINE Guide Inquiry Activities with confidence by watching the Inquiry Activity Teacher Preview video as you plan. After students complete the activity, the Inqu... missed class, a...

ADDITIONAL RES...
Inquiry Activity Teacher Preview

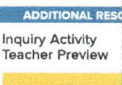

Teacher N...

Blue materials are included in the Collaboration Kits.

Module: Information Processing and Transfer

Inquiry Activity Planner

In this module, students will investigate information processing and transfer and design and build a device that uses light and/or sound to communicate a message.

Lesson	Inquiry Activity		Materials	
	★ GO ONLINE for teacher support videos on selected activities. Materials included in the Collaboration Kit are listed in blue.		Consumable	Non-Consumable
Lesson 1	**Hands On** Sense of Touch	30 min, small groups	material for blindfold	3 sandpaper samples of different grades, hand lens
	Purpose: To explore how their sense of touch works when their sense of sight is impaired.			
	Hands On Pill Bugs	30 min, small groups	15 pill bugs, potting soil, leaves, paper towels, water, fish food	hand lens, plastic habitat
	Purpose: To investigate how pill bugs use their senses to help them survive.			
Lesson 2	**Hands On** How Light Travels	30 min, small groups	white paper, batteries, clear cup, cup, water, index card	mirror, flashlight, protractor, sand, hand lens
	Purpose: To investigate how light travels and what types of objects reflect light.			
	Hands On It's Time to Focus	30 min, pairs	sheet of white paper	hand lens, desk lamp (teacher use only), various desk items: stapler, mug, tape dispenser (teacher use only)
	Purpose: To make a model to show how an animal eye works to refract light, and investigate what happens when the distance between the lens and retina is changed in a model eye.			
Lesson 3	**Hands On** Secret Message	30 min, small groups	batteries	flashlight
	Purpose: To investigate how patterns are used to transfer information.			
	Hands On Morse Code Message	30 min, pairs	batteries	flashlight, classroom objects
	Purpose: To use Morse code to send a message.			
	Research What's That Say?	30 min, pairs		
	Purpose: To research and decode a binary code message.			
STEM Module Project	**Engineering Challenge** Pixel Message	30 min, pairs	batteries	stopwatch, flashlight, bell, whistle, drum, translucent colored sheets, 2 colors
	Purpose: To design and build a device that uses light and/or sound to send a message across a room.			

McGraw-Hill is your partner for hands-on materials. To order new Collaboration Kits or refill specific items, contact the McGraw-Hill Education customer support line at (800) 336-3987.

Phenomena-Driven, Inquiry-Based, Hands-On Learning

Inspire All Students

Differentiation and ELD Support

Inspire Science has been designed to ensure that ALL students have access to quality, intellectually-rich science and engineering curriculum that supports language development and provides engaging learning opportunities. Here's how!

Uniting Phenomena

Phenomenon-driven instruction levels the playing field for learners by allowing them to access the core science content through a shared experience, observing a highly relevant real-world phenomenon. When students feel a personal connection to the phenomenon, they are more invested in aggregating the knowledge needed to explain the event. It is through these shared occurrences and supported instruction that learning is truly accessible to ALL students as they work toward achieving their learning goals.

Differentiated Instruction

Inspire Science incorporates the research-based Universal Design Learning Principles to provide educational practices that support multiple means of engagement, representation, action, and expression to ensure that all students have access to rigorous curriculum.

Robust differentiation support is found within the Teacher's Edition, as well as through leveled informational text resources, such as the Leveled Readers and Investigator articles. Support with practical strategies is found at the module and lesson level at multiple points. Leveled text aligns with the lexile ranges of the CCSS.

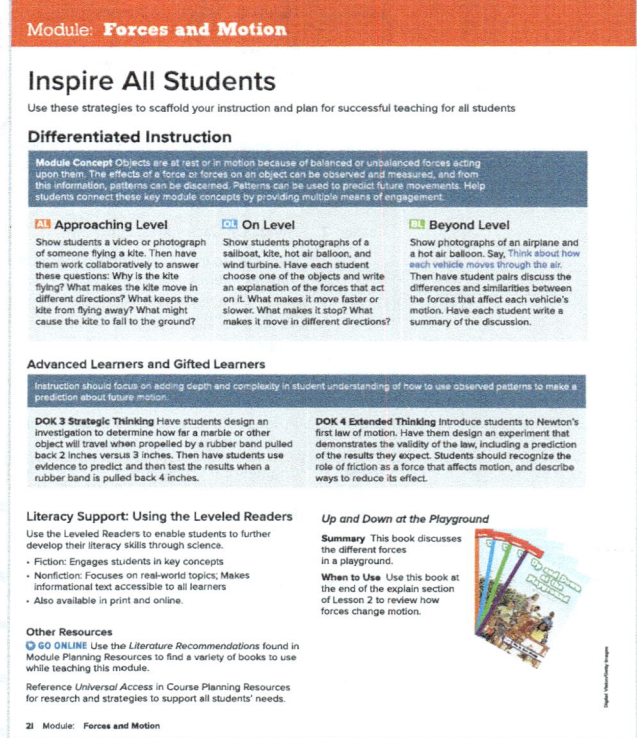

English Language Support

Home Language Support Build on and make use of students' home language to support their science learning in English. Teach students how to identify and use cognates to create linguistic bridges between school science and home to capitalize on emerging bilingualism.

EMERGING
Cognate Strategies Demonstrate the meaning of cognates by writing the word animal on the board. Ask students to tell you what the word means using words, phrases or gestures. Say and point to the word animal and have students repeat. Then have students say the word in their home language. Guide students to notice that the pronunciation is a little different but the spelling is not different. Write animals and animales on the board. Guide students to notice the differences in spelling and pronunciation in the plural form. There are many cognates in this module. Ask students to keep a list of the words they see that are similar in their home language.

EXPANDING
Cognate Strategies Explain the meaning of cognates by writing the words animals and animales on the board. Ask students to tell you the meaning of the words. Then support students in finding the differences and similarities in sounds and letters. For example, both words have the same spelling except that one ends in s and the other in es. Say the word animals and have students say animales. Note that there is not a lot of difference in spelling or pronunciation. There are many cognates in this module. Encourage students to list the cognates, noting the differences in spelling and pronunciation as you work through the module.

BRIDGING
Cognate Strategies Ask students to tell you if they know what a cognate is, i.e. a word that looks similar, sounds similar, and shares a meaning across some languages. Have students read the title of the module to find the cognate, animal. Have them tell you the word in their home language (animal) and give you a definition of the word in English. Point out that the plural animals/animales have different spellings. Throughout the module, students will find many cognates. When beginning a new page, ask students to scan the page for cognates and add them to a list along with their definitions in English.

English Language Support

Inspire Science applies the best instructional practices for teaching EL students. Each module and lesson has scaffolded activities that offer students of any level of English language proficiency the opportunity to engage in academically challenging science and engineering content while supporting language acquisition.

Strategies and activities allow for EL instruction specific to each of your students.

Language Building Resources

Inspire Science lessons carefully integrate reading, writing, speaking, listening, and collaborating into each lesson. This structure provides EL students purposeful language usage and resource access to convey their understanding in a meaningful way.

Cognates
Cognates are words in two different languages that share a similar meaning, spelling, and pronunciation. Review differences in spelling and pronunciation of these terms with your Spanish-speaking English Learners.

mammal mamífero	**insect** insecto	**reptile** reptil
amphibian anfibio	**protection** protección	**signal** señal
armadillo armadillo	**zebra** cebra	**lion** león

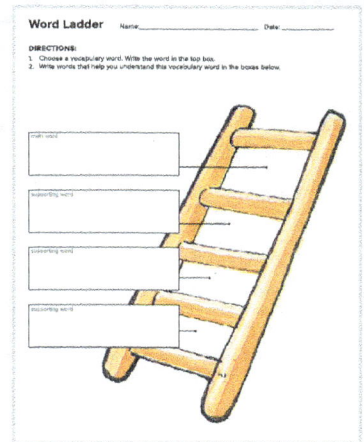

Advanced Learners and Gifted Learners

Provide your advanced and gifted learners with challenging activities that identify the Depth of Knowledge (DOK) to provide enrichment opportunities for demonstrating advanced performance in science and engineering. This is in addition to the Approaching Level, On Level, and Beyond Level support included in the differentiated instruction strategies for each module and lesson.

Cross-Curricular Connections

Inspire Science was built to help students develop language and mathematics skills in ways that support learning science and engineering while also supporting ELA and math goals. Every lesson integrates cross-curricular connections to the Common Core State Standards for ELA/Literacy and Common Core State Standards for Mathematics in alignment with the NGSS.

CROSS-CURRICULAR ▸ Connections

Math Integration

Science and math are closely related in the real world—a key reason for the Science and Engineering Practice of Using Mathematical and Computational Thinking, as well as Analyzing and Interpreting Data. In *Inspire Science*, students will engage with math the same way that real scientists and engineers do. They will collect and analyze data, create graphs, and make connections between mathematics and real-world events to solve challenging problems.

INQUIRY ACTIVITY

Hands On

Forces Affect the Way Objects Move

You saw people going down a slide. A slide is one kind of ramp. Investigate how the height of a ramp will change a toy car's motion.

Make a Prediction How will the height of a ramp affect the motion of a toy car?

Materials
- 4 books
- cardboard
- masking tape
- toy car
- meterstick

Carry Out an Investigation

1. Stack two books on the floor. Lean a piece of cardboard along the top book to make a ramp. Tape the edge of the cardboard to the floor.
2. Place a toy car at the top of the ramp. Release the car.
3. **MATH Connection** Use the meterstick to measure the distance the car traveled.
4. **Record Data** Record the distance the car traveled in the data table.
5. Repeat steps 2–4 for a total of three trials.
6. Repeat steps 1–5 with a stack of four books.

Distance Traveled in Centimeters

	Trial 1	Trial 2	Trial 3
Two-book ramp			
Four-book ramp			

7. Compare the distances the toy car traveled with the two ramps. What pattern do you see?

8. Predict what would happen if your ramp had six books.

MATH ▸ Connection

Math connections are found in relevant places within the modules, including the inclusion of practical math skills within the inquiry activities.

Inspire Science

Literacy Integration

Integrating literacy with your science instruction will help your students build literacy skills while learning science. By incorporating our leveled, nonfiction reading content, you will see your students' close reading and communication skills improve with text-dependent questions, paired readings, arguments, narratives, and collaborative conversations practiced in the context of science that's fun!

Science Literacy Framework

The CER Framework helps students construct explanations for science phenomenon using evidence they have gathered throughout the module.

CLAIM
↓
EVIDENCE
↓
REASONING

Close Reading Framework

The Close Reading activities in Explain guide students to search for answers to text-dependent questions within informational text passages.

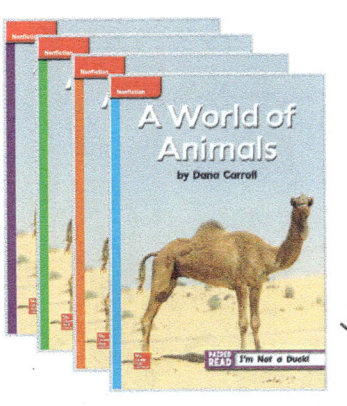

Leveled Readers

Every module includes a leveled reader title that is available in four levels.

-  Approaching Level
- On Level *Available in Spanish*
-  Beyond Level
- EL

Investigator

These books provide a collection of engaging articles about real-world science and engineering stories, available in two levels.

Available in Spanish

- Approaching Level (online, printable)
- On Level

Primary Sources

Use primary sources to learn about scientists and engineers and their fascinating discoveries.

Cross-Curricular Connections 23

STEM Connections

While career opportunities in Science, Technology, Engineering, and Math (STEM) increase each year, qualified candidates for these careers continue to fall short. This is known as the *STEM Gap*. This gap represents a great opportunity for the students in your classroom today. The real-world STEM Connections and the avatar-based STEM Career Kids in *Inspire Science* will help your students imagine a career they might like to pursue some day—a key factor of student engagement. The wide variety of connections, whether real-world or avatar-based, represents a broad range of STEM careers, from jobs that require a high-school education to those that require a PhD.

Real-World STEM Connections

Inspire Science integrates real-world STEM Connections into each module and lesson with real-world scientists and engineers.

Microbial Ecologist

24 Program Design

Inspire Science

STEM Career Kids

In Grades K–4, the STEM Career Kid avatars provide an approachable and engaging introduction to STEM Careers for young learners.

STEM Module Project

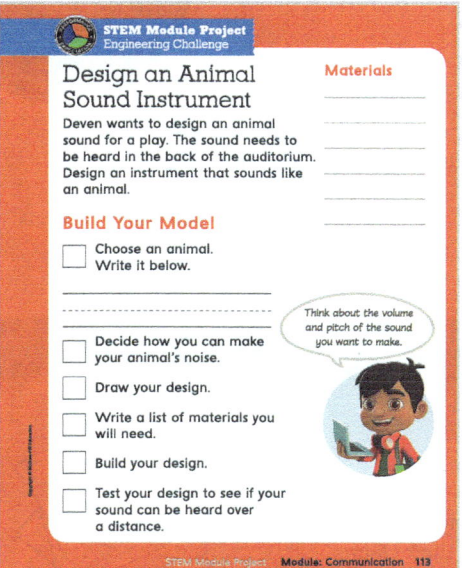

Module Wrap-Up

STEM Connection

The STEM Career Kids capture the imaginations of young learners.

MAYA
Geologist

STEM Connections 25

Next Generation Assessment Strategies

Three-Dimensional Learning Requires Three-Dimensional Assessments!

Inspire Science includes a variety of assessment options to support teachers with differentiation strategies and support students on their journey to mastery of the Performance Expectations.

Each *Inspire Science* lesson begins with a Formative Assessment Science Probe.

Formative Assessment

Formative assessment, embedded at many points throughout each module and lesson, facilitates student reflection on their thinking (metacognition) and allows teachers to dynamically differentiate instruction. Following are the types of formative assessment resources in *Inspire Science*, which you'll find online and in the print Student Editions.

PAGE KEELEY, M.Ed.

Page Keeley's Science Probes present the lesson phenomenon in an engaging way, promoting student thinking and discussion and revealing commonly-held preconceptions students bring to their learning to guide differentiated instruction strategies.

FEATURE	INSTRUCTIONAL PURPOSE
Science Probes	Found at the beginning of each lesson, **Science Probes** reveal student preconceptions to guide instruction.
Claim-Evidence-Reasoning	With the **CER Framework** (Claim, Evidence, Reasoning), found in certain lessons, students will make claims and document their reasoning during Explore and add evidence and revise their claims as needed later in the lesson.
Three-Dimensional Thinking Questions	Throughout each lesson, students will encounter questions that address at least two of the three dimensions of the NGSS to check progress with the SEPs, DCIs, CCCs, and Performance Expectations.
Talk About It	Throughout each lesson, student-initiated or teacher-led **Talk About It** prompts encourage discussion, allowing students to demonstrate their understanding of the phenomena, DCIs, or CCCs.
Inquiry Activities	In each inquiry activity (2–3 per lesson), students will encounter formative assessment questions that help bolster three-dimensional thinking.
Module Pretest	The **Module Pretests**, found at the beginning of each module in Grades 2–5, assess prerequisite knowledge of Disciplinary Core Ideas from prior grades to evaluate student readiness for the module.

Chris Keeley Photography

26 Program Design

Summative Assessment

Summative assessment tools at the module and lesson level help ensure lasting learning and alignment of student skills to the Performance Expectations. Following are the summative assessment tools found in *Inspire Science*, both online and in the print Student Editions.

FEATURE	INSTRUCTIONAL PURPOSE
Three-Dimensional Thinking Questions	At the end of the lessons, students will demonstrate their understanding of at least two of the three dimensions of NGSS to develop three-dimensional thinking skills.
Lesson Check	Found in every lesson online, **Lesson Checks** determine how students are building a progression of learning toward the Performance Expectations.
Module Test	Found at the end of each module online, **Module Tests** evaluate student proficiency against the Performance Expectations with multiple choice, extended response, constructed response, and performance-task items.
STEM Module Project Performance-Based Rubrics	With each STEM Module Project, found at the end of each module, students will complete **Performance-Based Rubrics** and answer summative questions to demonstrate how they've applied their knowledge and understanding of the Performance Expectations to their project.
Vocabulary Check	Through online interactives, students practice and check their understanding of science language. Immediate feedback from the system is provided.

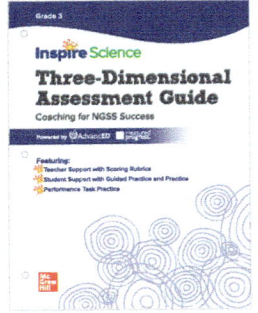

Three-Dimensional Assessment Guide

Through completing the Inspire Science units and practicing with the discrete items and performance tasks your students will have a feeling of confidence going into the year-end assessment.

The *Inspire Science* Three-Dimensional Assessment guide, available in print and digital formats, provides Guided Practice and Practice items that will prepare students for success.

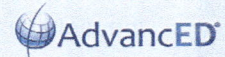 

The NGSS call for higher rigor, a greater demand for evidence of student learning, and the integration of the three dimensions (Science and Engineering Practices, Disciplinary Core Ideas, and Crosscutting Concepts). Integrated throughout the *Inspire Science* assessments are AdvancED/Measured Progress® assessment items that provide students with opportunities to demonstrate an understanding of the three dimensions of the Performance Expectations as instruction occurs.

Professional Learning

We know it can be a challenge to implement a new science program with new standards. That's why *Inspire Science* comes with a library of relevant, self-paced, professional learning videos and modules to support you from implementation through instructional progression and mastery, all available 24/7, at your fingertips.

Program Implementation Support

Implementation support provides everything you need to know to get up to speed on the first day of school.

- **Quick Start eLearning Modules** explain program basics to help get you started.
- **Plan, Teach, and Assess eLearning Modules** provide deep-dives of the program's instructional model and resources.

Digital Platform Support

In the Technical Support Resource Library, you will find step-by-step instructions for each of your digital tools to help you feel confident planning, teaching, and assessing in the digital experience.

Inspire Science

Ongoing Pedagogy Support

With *Inspire Science*, you will find a wide range of resources on key instructional and pedagogical topics, including videos from our program authors and consultants.

- **STEM Classroom Videos** model lessons from real classrooms
- **Science Preconceptions Videos** review common preconceptions and strategies to overcome them
- **Instructional Coaching Videos** discuss best practice strategies and the "Why" behind the success
- **Teacher Activity Videos** show planning tips and expected results to help with hands-on activity time
- **Science Pedagogy Micro-Courses** provide facilitation guides for both self-guided or small-group courses

Finding Your Professional Learning Resources

All professional learning resources are easily identifiable in your digital experience. Just look for the apple icon in your course, module, or lesson pages.

Authors and Partners

Program Authors

Dr. Doug Fisher
Dr. Douglas Fisher is Professor of Educational Leadership at San Diego State University and a teacher leader at Health Sciences High & Middle College. He is a member of the California Reading Hall of Fame and recipient of many awards for excellence in education. He has published numerous books and articles and is the co-author of *Visible Learning for Science, Grades K-12* and *Reading and Writing in Science: Tools to Develop Disciplinary Literacy*. He is also an ASCD author, keynote presenter, and President of the International Reading Association.

Dr. Jay Hackett
Dr. Jay Hackett is an emeritus professor of Earth Sciences and past recipient of the William R. Ross Science Award as an Honored Alumnus at the University of Northern Colorado. Dr. Hackett is co-author of *Teaching Science as Investigations* and made contributions to the development of *Inquiry and the National Science Education Standards: A Guide for Teaching and Learning*. Dr. Hackett is an admired science educator and McGraw-Hill Education science author.

Page Keeley, M.Ed.
Page Keeley, M.Ed., is a nationally-renowned expert on science formative assessment and teaching for conceptual change. She is the author of several award-winning books and journal articles on uncovering student thinking using formative assessment probes and techniques. She was the Science Program Director at the Maine Mathematics and Science Alliance for 16 years and a past President of the National Science Teachers Association. Currently she is an independent consultant providing professional development to school districts and science education organizations and a frequent invited speaker at national conferences.

Dr. Jo Anne Vasquez
Dr. Jo Anne Vasquez, a past President of the National Science Teachers Association and the National Science Education Leadership Association, was the first elementary educator to become a Presidential Appointee to the National Science Board, the governing board of the National Science Foundation. Her distinguished service and extraordinary contributions to the advancement of science and STEM education at the local, state, and national levels has won her numerous awards: 2014 National Science Education Leadership Award for Outstanding Leadership in Science Education, 2013 National Science Board Public Service Award, and "Robert H. Carlton Award" for Leadership in Science Education.

Dr. Richard Moyer
Dr. Richard Moyer is an emeritus professor of Science Education and Natural Sciences at the University of Michigan-Dearborn. He is an award-winning educator, author, and co-author of *Everyday Engineering: Putting the E in STEM Teaching and Learning, Teaching Science as Investigations,* and *More Everyday Engineering*. Dr. Moyer has served for more than 33 years as a McGraw-Hill Education science author.

In Memoriam Dr. Dorothy J.T. Terman
Dr. Dorothy J.T. Terman served for 21 years as Science Coordinator for California's Irvine Unified School District, where she was responsible for science curriculum development, program implementation, and assessment. She held a B.S. in Science Education from Cornell University, an M.A. in Cell Biology from Columbia University, and a Ph.D. in Curriculum from the University of Iowa. She received many awards, including the Ohaus Award from the National Science Teachers Association for Innovation in Elementary Science Education. She was a consultant for inquiry-based science curriculum implementation and a veteran McGraw-Hill Education science author. We will miss her inspiration and passion for science education

Dinah Zike, M.Ed.
Dinah Zike, M.Ed. is an award-winning author, educator, and inventor known for designing three-dimensional hands-on manipulatives and graphic organizers known as Foldables® and VKVs® (Visual Kinesthetic Vocabulary®). Ms. Zike is the founder and President of Dinah-Might Adventures, LP and Dinah Zike Academy. She is also the recipient of the Teachers' Choice Award For the Classroom and Teachers' Choice Award For Professional Development.

Key Partners

The Concord Consortium is a nonprofit educational research and digital learning organization focused on delivering the promise of technology for education in science, math, and engineering. The *Inspire Science* simulations, created in partnership with The Concord Consortium, enable students to model concepts otherwise not possible to explore in the classroom.

Filament Games creates digital learning games and interactives designed to foster 21st-century skills through experiential learning. The immersive games included with *Inspire Science*, developed in partnership with Filament Games, enable students to "play" with the lesson concepts to deepen conceptual understanding.

Measured Progress, a not-for-profit organization, is a pioneer in authentic, standards-based assessments. Included with *Inspire Science* is **Measured Progress STEM Gauge®** assessment content, which enables teachers to monitor progress toward learning NGSS.

Program Advisors

Phil Lafontaine
NGSS Educational Consultant
Sacramento, California

Emily C. Miller
University of Wisconsin at Madison
Madison, Wisconsin

Dr. Timothy Shanahan
Distinguished Professor Emeritus
University of Illinois at Chicago
Chicago, Illinois

Jody Skidmore Sherriff
Regional Director
K-12 Alliance/WestEd
Sacramento, California

Content Reviewers

Jennifer Covarrubias
Teacher
Kinetic Academy
Huntington Beach, California

Monica Galavan, M.A., M.S
Cajon Valley Union School District
Teacher
San Diego, California

Mika L. George, B.S.
Highland Local School District
Teacher
Sparta, Ohio

Teresa Harris-Belcher, B.A.
Highland Local School District
Teacher
Sparta, Ohio

Dr. Cindy Klevickis
James Madison University
Professor of Integrated Science and Technology
Harrisonburg, Virginia

Amy Syverson Kunis, M.A.
Perris Elementary School District
Teacher
Perris, California

Kathi Lundstrom, Ph.D.
Norwalk-La Mirada Unified School District
Teacher
La Mirada, California

Lisa K. Reely
Highland Local School District
Teacher
Sparta, Ohio

Derrick Svelnys, M.S., M.Ed.
Chicago Public Schools
Teacher
Chicago, Illinois

Tasha Terrill, M.Ed.
Highland Local School District
Teacher
Sparta, Ohio

Amanda Waggoner
Highland Local School District
Teacher
Sparta, Ohio

Kimberly Wilson, M.Ed.
Mariners Christian School
Teacher
Costa Mesa, California

Kandi K. Wojtysiak, M.Ed.
Notre Dame Preparatory High School
Science Department Chair
Scottsdale, Arizona

Module and Lesson Structure

Module and Lesson Walk Through

This section will provide you with a step-by-step tour of a module. Become familiar with the print and digital activities and resources available in each module of *Inspire Science*. Here you will find examples of the following:

* Correlations for the NGSS Performance Expectations
* Module and Lesson Planning Resources
* Module Opener
* STEM Module Project Launch
* 5E Lesson
* STEM Module Project
* Module Wrap-Up

> **Need login credentials?**
> Go to my.mheducation.com and select "Create Teacher Account."

Module and Lesson Planning Resources

The *Inspire Science* Teacher's Edition provides easy-to-follow correlations to the Next Generation Science Standards, telling you which modules address which Performance Expectation.

Performance Expectations and NGSS Correlations

At the beginning of each unit, correlations show how the modules within the unit align to the NGSS in the **Performance Expectations at a Glance** feature. This table identifies where students will discover and practice the Science and Engineering Practices, Disciplinary Core Ideas, and Crosscutting Concepts needed to succeed with each Performance Expectation. Every module clearly identifies by page number the *Inspire Science* resources that correlate to the Next Generation Science Standards.

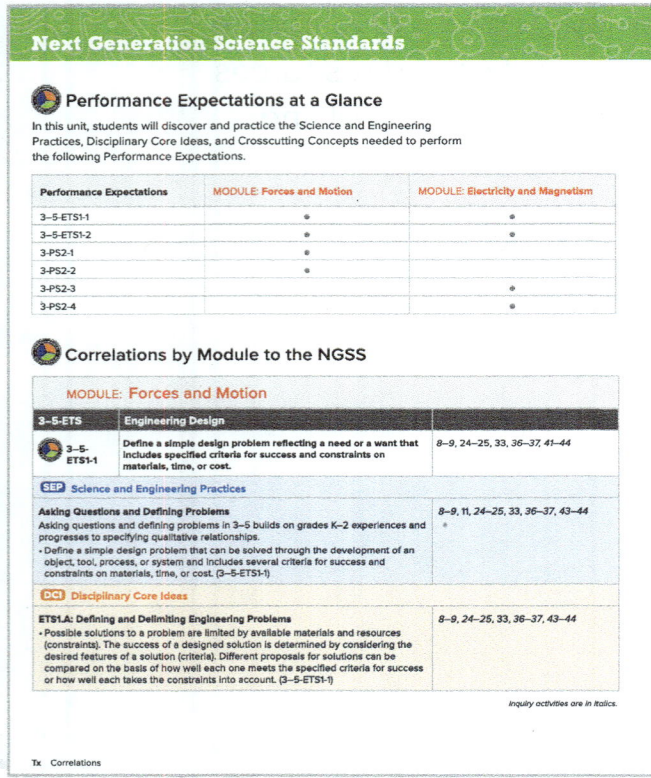
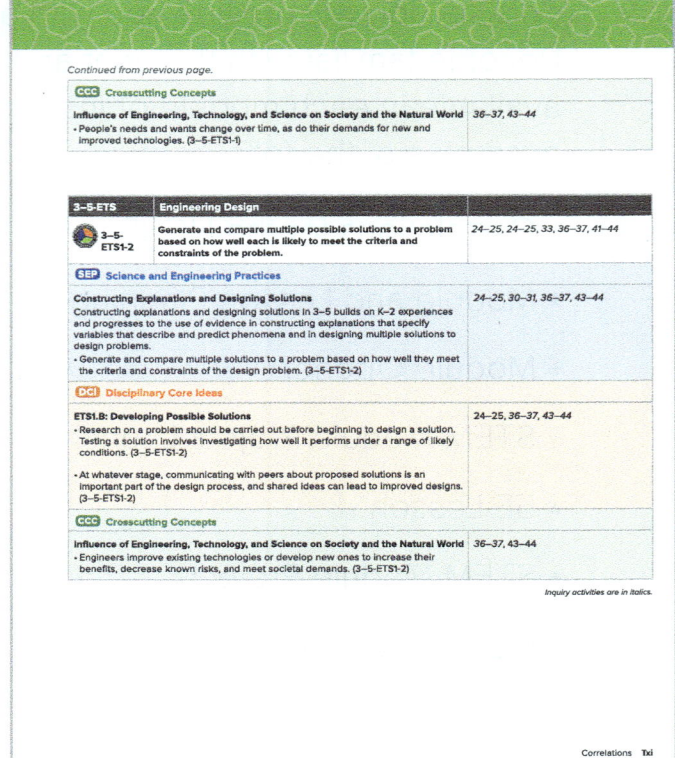

33 Module and Lesson Walk Through

Three-Dimensional Learning

Each module shows the three dimensions of learning that enable students to achieve proficiency with the Performance Expectations addressed in the module.

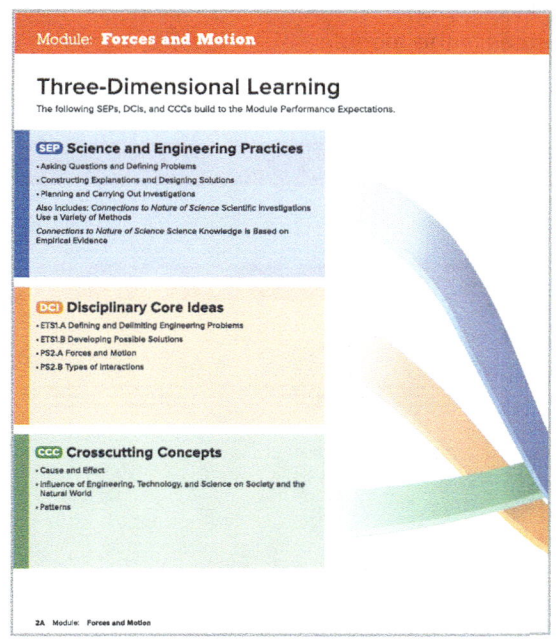

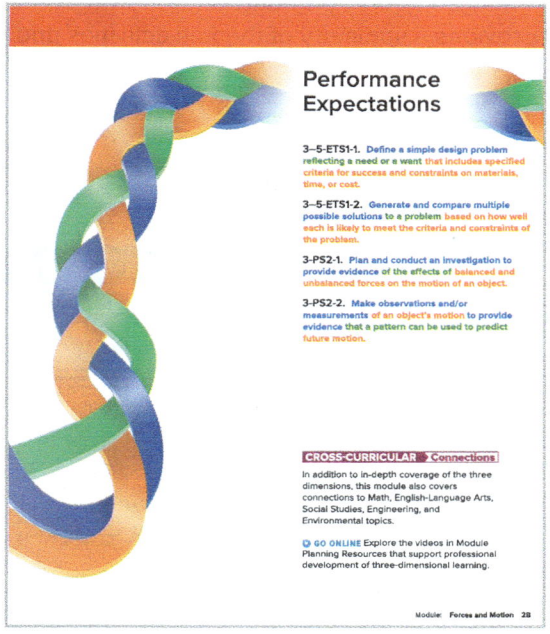

Disciplinary Core Idea Progression

This table illustrates in detail the Disciplinary Core Idea Progressions across grades K–8.

Three Dimensions at a Glance

Use this chart to locate where students will encounter each of the three dimensions that build to the Performance Expectations in the module.

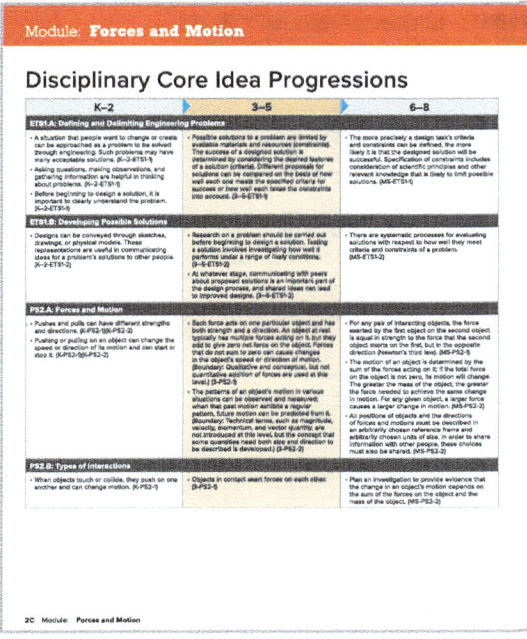

Module and Lesson Planning Resources 34

Module and Lesson Planning Resources

The Module and Lesson Planner pages provide a high-level look at what students will use to learn and master the Performance Expectations.

Module Planner

The **Module Planner** provides a summary of the key activities and resources in the module as well as pacing recommendations.

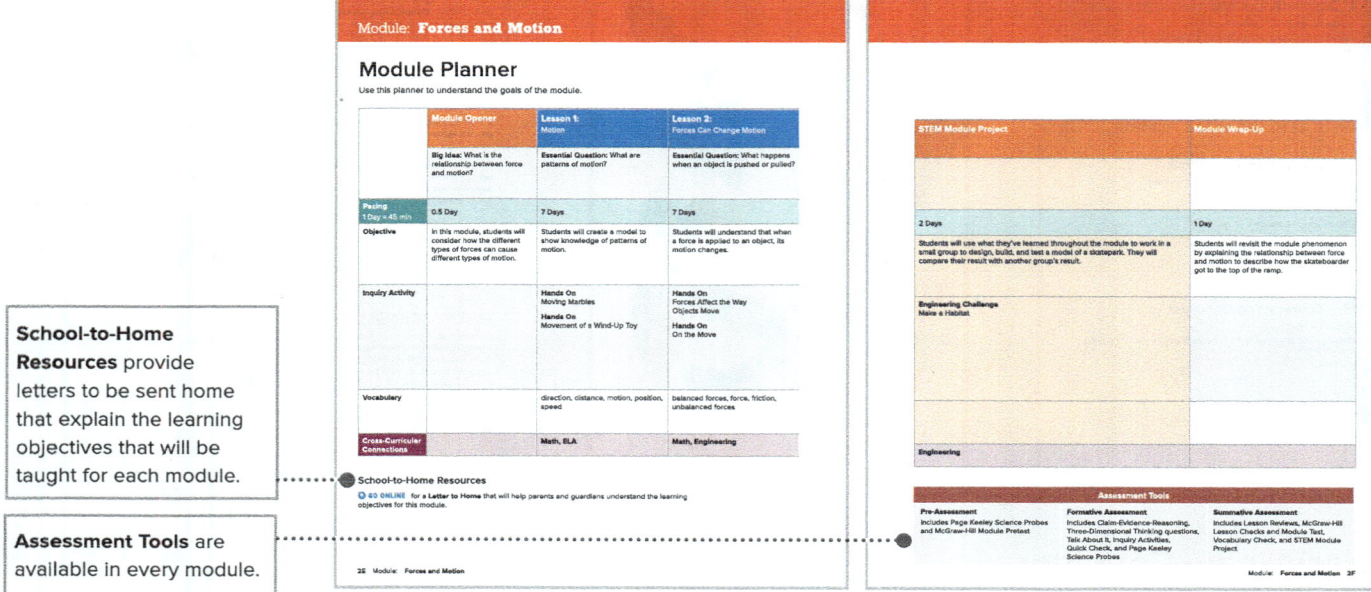

School-to-Home Resources provide letters to be sent home that explain the learning objectives that will be taught for each module.

Assessment Tools are available in every module.

Inquiry Activity Planner

The **Inquiry Activity Planner** helps you get ready for all inquiry activities in the module, with a summary of the activity, the purpose, pacing and grouping strategies, and needed materials.

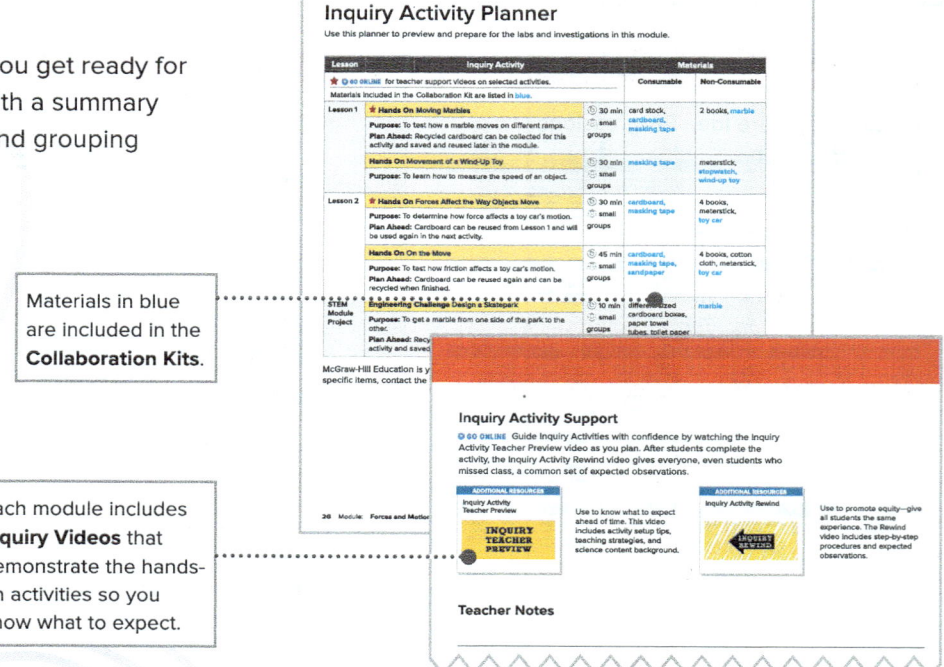

Materials in blue are included in the **Collaboration Kits**.

Each module includes **Inquiry Videos** that demonstrate the hands-on activities so you know what to expect.

35 Module and Lesson Walk Through

Inspire All Students

Each module includes strategies to scaffold instruction and plan for successful teaching for all students.

Differentiated Instruction strategies suggest leveled activities for Approaching Level, On Level, Beyond Level, and Advanced and Gifted Learners.

English-Language Support provides suggested strategies and activities for EL students in alignment with the EL Framework (Emerging, Expanding, Bridging).

Add depth and complexity for your **Advanced and Gifted Learners.**

Literacy Support helps students to further develop close reading skills through science.

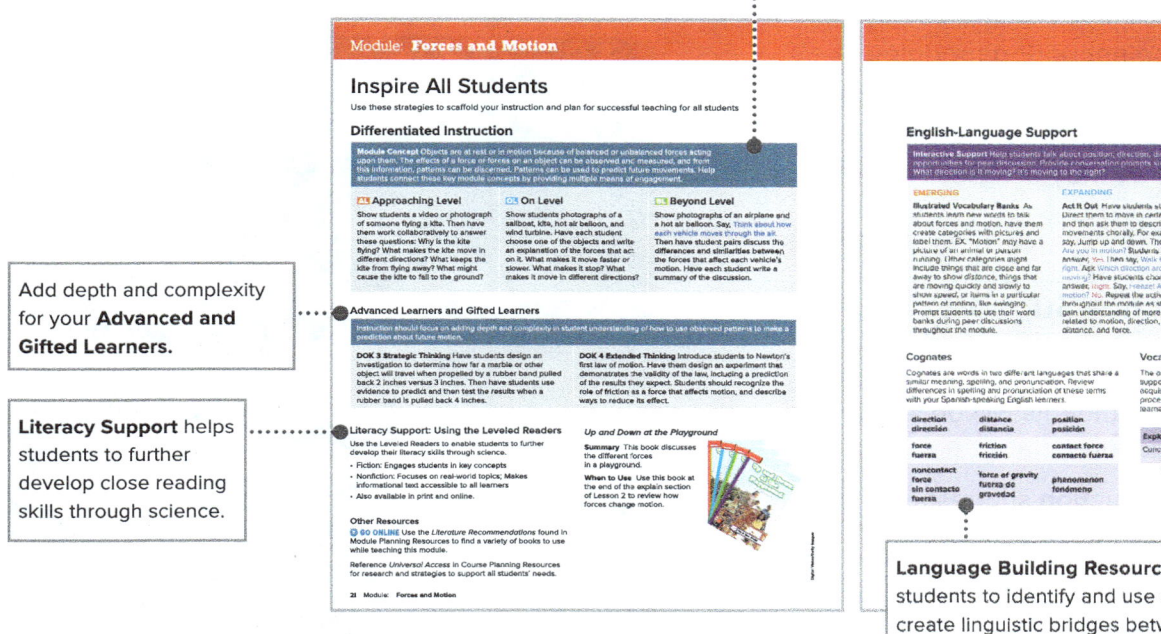

Language Building Resources teach students to identify and use cognates to create linguistic bridges between school and home to capitalize on emerging bilingualism.

Lesson Planner

Building to the Performance Expectations details the three dimensions of learning that your students will explore to develop mastery of Performance Expectations.

The **Track Your Progress** table helps you monitor student progress toward mastery of the Performance Expectations.

Every lesson offers three pacing options to best meet your schedule.

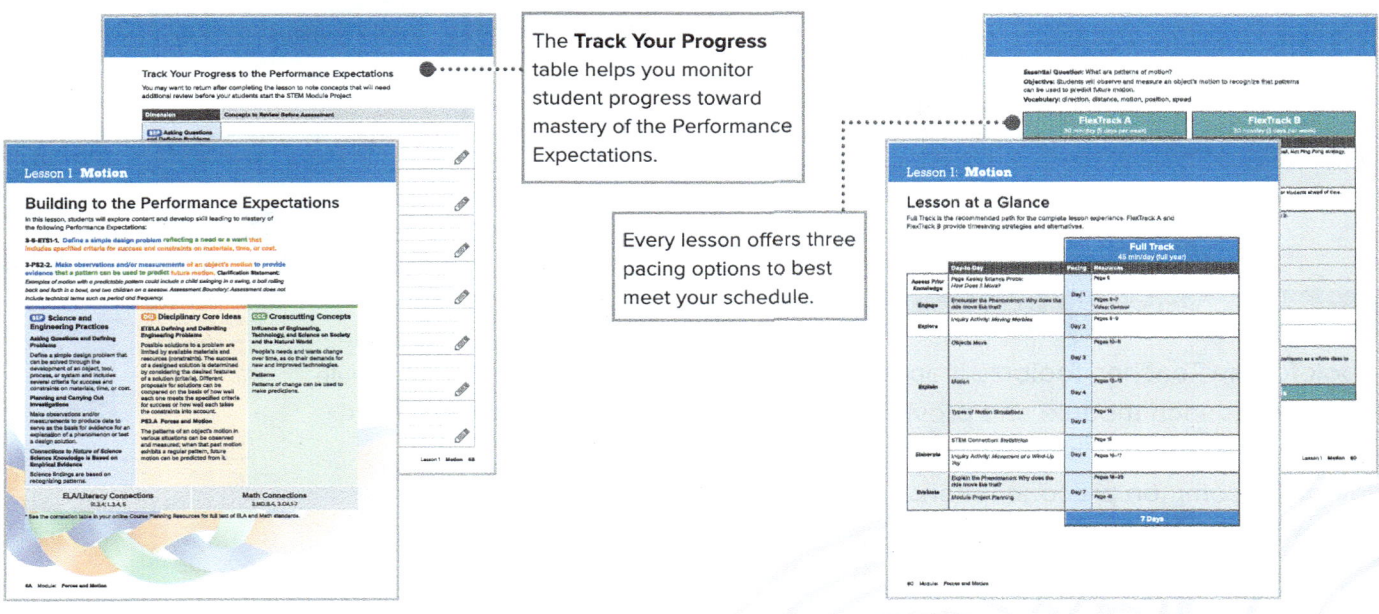

Module and Lesson Planning Resources 36

Module Opener

MODULE OPENER • STEM MODULE PROJECT LAUNCH • LESSON LAUNCH SCIENCE PROBE • ENGAGE • EXPLORE

Module Opener
AT-A-GLANCE

Inspire your students' curiosity with a real-world phenomenon that inspires students to ask questions and investigate the world around them. The anchoring module phenomenon will uncover students' initial ideas, setting them up to see how their thinking evolves as they progress through the module.

Inspiring Teacher Support

Performance Expectations are identified to let you know what students will be learning throughout the module.

Differentiated Instruction suggestions help you provide instruction that is just right for students of all levels.

Word Walls are included for students in Grades K–1 to emphasize key foundational vocabulary.

Teacher Toolbox

Look for the Teacher Toolbox. It appears throughout each module and provides science background information or to or identifies common preconceptions related to the content at hand.

STEM Connections

Real-world STEM Careers (with relatable STEM Career Kids in K–1) are introduced at the module level to help students see how the information from the module is applied in the real world.

37   Module and Lesson Walk Through

Inspire Science

EXPLAIN · ELABORATE · EVALUATE · MODULE PROJECT PLANNING · MODULE PROJECT COMPLETION AND MODULE WRAP-UP

ENCOUNTER THE PHENOMENON

The **Module Opener** begins the inquiry process by presenting an anchoring phenomenon to explore throughout module. Lesson-level investigative phenomena and inquiry activities help students build understanding of the module phenomenon.

GO ONLINE

Go Online to Explore

Interactive digital content gets students thinking and talking about the module phenomenon.

Talk About It

In each **Module Opener**, students are prompted to discuss the module phenomenon after reviewing the **ENCOUNTER THE PHENOMENON** resource.

Did You Know?

Did You Know statements provide background information to promote conversation and help students turn their observations into questions they will answer later.

Module Opener 38

STEM Module Project Launch

MODULE OPENER | STEM MODULE PROJECT LAUNCH | LESSON LAUNCH SCIENCE PROBE | ENGAGE | EXPLORE

STEM Module Project Launch
AT-A-GLANCE

In grades 2 and up, build excitement and get your students curious about what they'll be learning in each lesson. This section tells students about the project they'll complete at the end of the module and how the lessons in the module will help them in their planning. Your students will start asking questions, setting goals, and preparing to experience the engineering design process like the professionals.

PHASE 1 (Grades 2-5)

STEM Module Project Launch
Engineering Challenge

Students assume the role of a scientist or engineer and are charged with the task of designing a solution to the related Science or Engineering Challenge at the end of the module.

As students progress through each lesson, they will generate questions and begin initial planning while learning about the related, real-world STEM Career.

EXPLAIN · ELABORATE · EVALUATE · MODULE PROJECT PLANNING · MODULE PROJECT COMPLETION AND MODULE WRAP-UP

PHASE 2 (Grades 2–5)

STEM Module Project Planning

After each lesson, students have the opportunity to think about how what they've just learned can help them with their project at the end of the module.

What do you think you need to know before you can design a skatepark?

PHASE 3 (Grades K–5)

STEM Module Project

At the end of the module, students will complete the Science or Engineering Challenge.

ENGINEERING CHALLENGE
In this STEM module project, students will follow the **Engineering Design Process** to design, construct, and test a skatepark.

SAM
Architectural Drafter

STEM Module Project Launch **40**

Lesson Launch / Science Probe

MODULE OPENER · STEM MODULE PROJECT LAUNCH · **LESSON LAUNCH SCIENCE PROBE** · ENGAGE · EXPLORE

Science Probe
AT-A-GLANCE

One of the most effective ways to support conceptual learning is through formative assessment. That is why *Inspire Science* begins every lesson with a formative assessment science probe to assess students' prior knowledge.

Science probes present a real-world phenomenon, or core concept, to promote student thinking and discussion, revealing commonly-held preconceptions and initial ideas students bring to their learning so you can best inform your instruction.

Inspiring Teacher Support

Detailed teacher support for every science probe:

- Research-based, common preconceptions associated with the content of the lesson
- Suggested Page Keeley discussion strategies and support videos
- Detailed account of the purpose and usefulness of each probe
- Clearly stated teaching and learning implications
- Scientific explanations to clarify the specific content at hand

Page Keeley **Productive Discussion Strategies** provide a variety of ways to get students talking and documenting their thinking. A strategy is recommended for each science probe including specific Page Keeley strategy videos.

PAGE KEELEY, M.Ed.
Author and Educator

Uncover Student Preconceptions

Page Keeley, M.Ed. is a nationally-renowned expert on science formative assessment and teaching for conceptual change. She is the author of several award-winning books and journal articles on uncovering student thinking using formative assessment probes and techniques. She was the Science Program Director at the Maine Mathematics and Science Alliance for 16 years and a past President of the National Science Teachers Association.

Sticky Bar Graph Productive Discussion Strategy

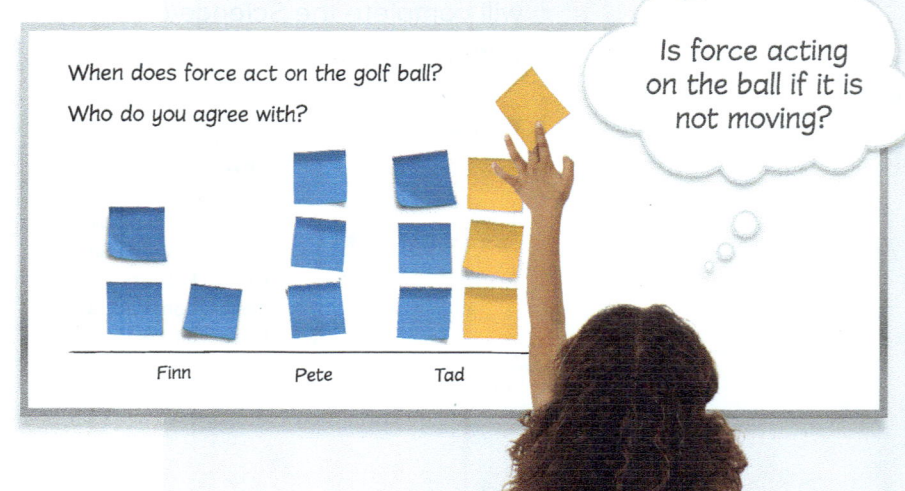

Inspire Science

EXPLAIN ELABORATE EVALUATE MODULE PROJECT PLANNING MODULE PROJECT COMPLETION AND MODULE WRAP-UP

Simple Illustration or Scenario

Science Probes present students with familiar real-world phenomenon or a core concept. These could be in the form of simple illustration or scenario.

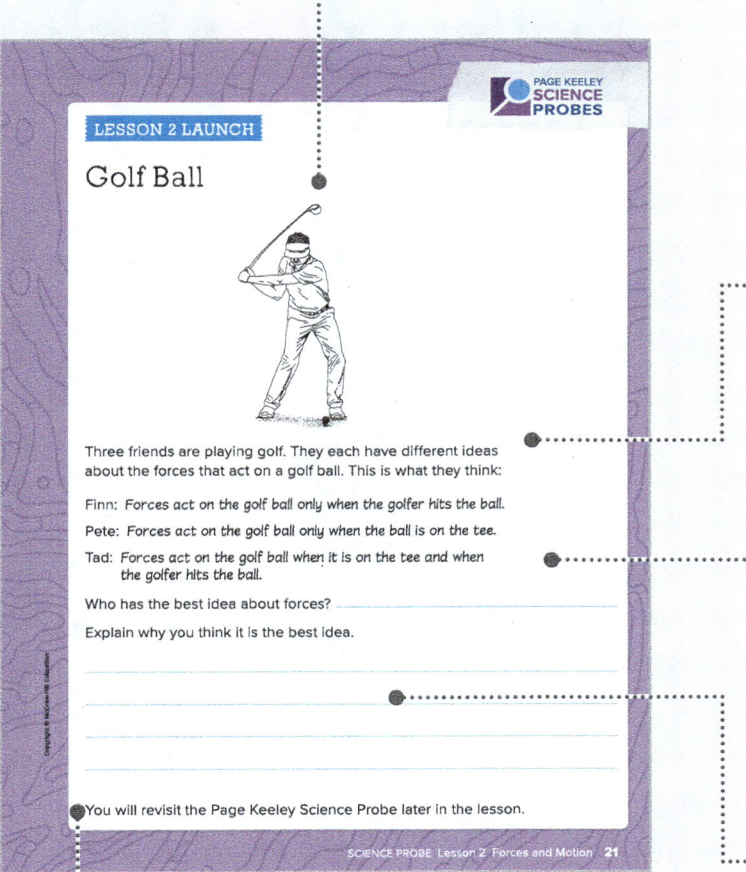

Real-World Phenomena

Relevant phenomena have great explanatory power. The situations presented are designed to draw out deeper thinking and elicit more thoughtful responses from students.

Best Versus Right Answer

Students are more motivated to learn in a non-judgmental environment. By referencing the "best answer" to explain thinking, rather than the "right answer," students feel safe in sharing their thinking.

Explanatory Answers Reveal Students' Thoughts

Students are required to provide an explanation for their answers, which helps uncover preconceived notions that may be clouding students' thought processes.

Revisit the Probe

Students will revisit the science probe throughout the lesson. After engaging with a variety of learning opportunities, students will be able to adjust their thinking if needed based on the evidence they've gathered in the lesson.

Lesson Launch / Science Probe

Engage

MODULE OPENER • STEM MODULE PROJECT LAUNCH • LESSON LAUNCH SCIENCE PROBE • ENGAGE • EXPLORE

Engage
AT-A-GLANCE

The Engage phase will inspire students' curiosity with a real-world phenomenon they will investigate throughout the lesson. These lesson phenomena help uncover student preconceptions and generate collaborative conversations that turn observations into questions to investigate.

As students progress through the lesson, they will begin to reveal answers to the questions they generated and will revisit their initial thinking to see how it changes as they learn new information.

Inspiring Teacher Support

Disciplinary Core Ideas and **Lesson Objectives** are clearly stated.

The **Encounter/Discover the Phenomenon** question is connected to the **Essential Question** for the lesson.

Discussion prompts are provided to help you facilitate collaborative conversations.

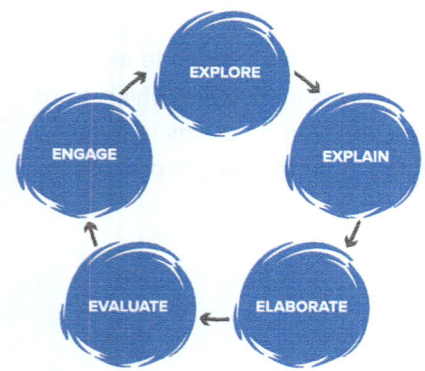

5E INSTRUCTIONAL MODEL

The 5E Instructional Model provides a proven, research-driven lesson flow with the flexibility to adjust as needed for your classroom needs.

 **Time To Move (Grades K–1)**

Engage younger students with suggested activities that get them up and moving.

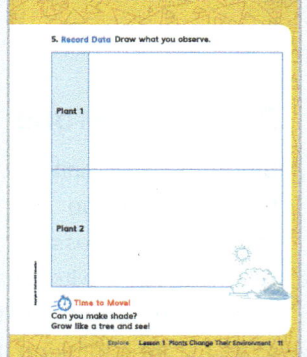

43 Module and Lesson Walk Through

Inspire Science

EXPLAIN · ELABORATE · EVALUATE · MODULE PROJECT PLANNING · MODULE PROJECT COMPLETION AND MODULE WRAP-UP

ENCOUNTER THE PHENOMENON

Students will engage with the lesson-level, investigative phenomena and collaborate to generate a list of questions.

ENCOUNTER THE PHENOMENON

How are they going down the slide so fast?

GO ONLINE
Check out *Slides* to see the phenomenon in action.

Talk About It
Look at the photo and watch the video of the kids going down the slide. What questions do you have about the phenomenon? Talk about your questions and observations with a partner.

Did You Know?
London has the longest and tallest slide in the world. It takes about 40 seconds to go down!

ENGAGE Lesson 2 Forces Can Change Motion 23

GO ONLINE

Go Online to Explore

Check out the video *Slides* to see the phenomenon in action.

Talk About It

Keep the Conversation Going

Students will describe what they see and turn their observations into questions that they will revisit and try to answer as they progress through the lesson.

Hifi Films/Shutterstock

Engage 44

Explore

MODULE OPENER — STEM MODULE PROJECT LAUNCH — LESSON LAUNCH SCIENCE PROBE — ENGAGE — **EXPLORE**

Explore
AT-A-GLANCE

The Explore phase lets your students get involved and investigate the phenomenon through a related, common experience. They will carry out an investigation, collect and interpret data, and begin to reveal answers to their questions and build understanding using different types of inquiry activities.

Inspiring Teacher Support

Inquiry activity support outlines the purpose, materials needed, and suggested strategies for facilitating the student work and discussions.

The **Science and Engineering Practices** are clearly highlighted, along with the **Crosscutting Concepts**, where relevant.

Differentiated Instruction tables provide activity customization suggestions to align with different levels of student skills.

Inquiry Spectrum provides flexible activity options to adjust the inquiry level to align with the learning needs of your students.

Engineering Connection activities are provided and include teacher support.

Collaboration Kits contain most of the materials needed for the hands-on inquiry activities. The materials are neatly organized and labeled to correlate with each unit and module, with enough materials for five groups of students.

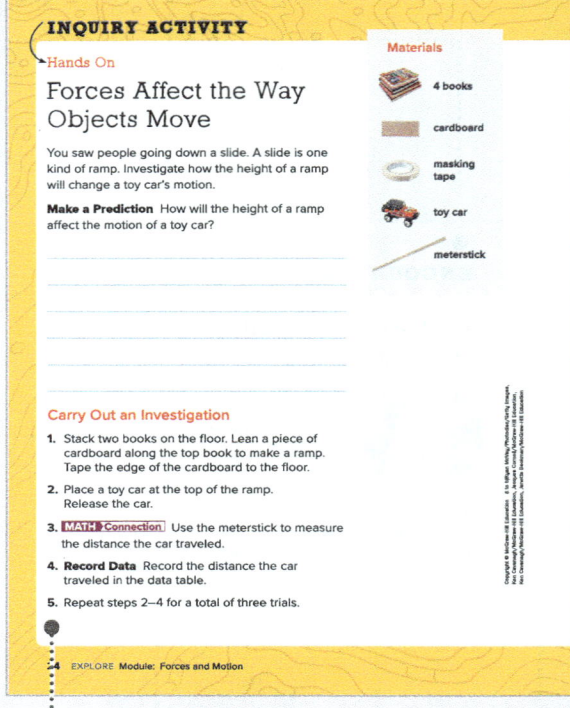

Inquiry activities guide students to think about the phenomenon, **make a prediction**, and **carry out an investigation** to test their predictions.

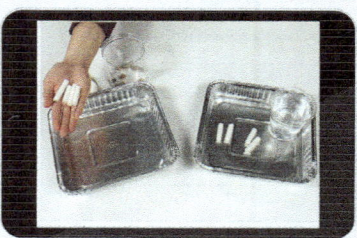

Inquiry Rewind
Inquiry activity videos provide a step-by-step look at the inquiry activity.

See the **Collaboration Kit Guide** for details regarding the materials that come in each kit.

Inspire Science

EXPLAIN · ELABORATE · EVALUATE · MODULE PROJECT PLANNING · MODULE PROJECT COMPLETION AND MODULE WRAP-UP

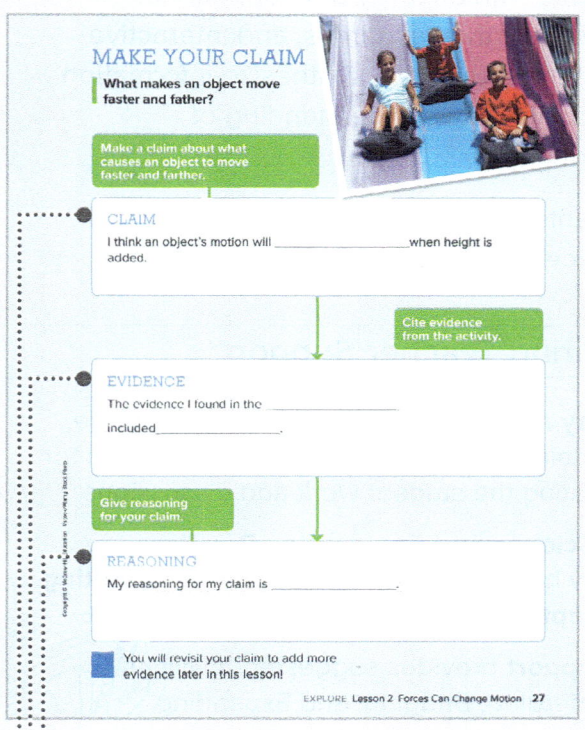

CLAIM EVIDENCE REASONING

Students will use the Claim, Evidence, Reasoning framework to help them as they explore and explain the phenomenon.

During their investigation, students will **record and analyze** their observations, think about changes to their prediction, and plan changes to their investigation.

At the end of the inquiry activities, students **communicate their findings** (with evidence) and **make connections to real-world examples**.

CLAIM Students reflect and brainstorm possible answers and take a clear stance on how the object will move.

EVIDENCE Students provide their initial evidence from what they learned in the inquiry activity. They return to their claim to add more evidence as it is revealed throughout the lesson.

REASONING Students explain the scientific knowledge, principle, or theory they used to support their argument.

Explore 46

Explain

MODULE OPENER · STEM MODULE PROJECT LAUNCH · LESSON LAUNCH SCIENCE PROBE · ENGAGE · EXPLORE

Explain
AT-A-GLANCE

This phase of the lesson model provides students with an array of informational text, supportive resources, and interactive activities so they can synthesize information and convey their understanding of the concepts.

Students will interact with the content and practice close-reading skills.

Inspiring Teacher Support

Inquiry activity support outlines the purpose, materials needed, and suggested strategies for facilitating the student work and discussions.

The **Science and Engineering Practices** are clearly highlighted, along with the **Crosscutting Concepts**, where relevant.

EL Support provides suggested activities for Emerging, Bridging, and Expanding student groups.

Differentiated Instruction tables provide activity customization suggestions to align with different levels of student skills.

Close Reading framework support to help you guide students through the Inspect, Find Evidence, and Make Connection steps.

Visual Literacy strategies and teacher support give students practice reading and understanding diagrams.

Vocabulary strategies encourage students to use context clues to derive the meaning of the vocabulary words.

Interactive Text

Students interact directly with core content to strengthen literacy and writing skills.

Leveled Reader

Students can extend their learning with leveled informational text that includes a paired fiction reading, text-dependent questions, hands-on activities, and graphic organizers to help summarize the selection.

- Approaching
- On Level
- Beyond
- ELL

47 Module and Lesson Walk Through

Inspire Science

EXPLAIN · ELABORATE · EVALUATE · MODULE PROJECT PLANNING · MODULE PROJECT COMPLETION AND MODULE WRAP-UP

Close Reading

Integrating literacy with science content helps students make connections while building close-reading skills and strengthening writing skills.

Access Complex Text

The ACT Framework (Access Complex Text) provides scaffolded practice for seven different complex text features.

Premade questions specific to the text help students understand complex text more clearly.

PRIMARY SOURCE

Students learn about scientists and engineers and their related discoveries through primary source features.

Crosscutting Concept Graphic Organizers

Use Crosscutting Concept Graphic Organizers to apply the themes to the science concept at hand throughout the lesson.

Explain 48

Elaborate

MODULE OPENER · STEM MODULE PROJECT LAUNCH · LESSON LAUNCH SCIENCE PROBE · ENGAGE · EXPLORE

Elaborate
AT-A-GLANCE

In Elaborate, students apply knowledge to new situations to develop a deeper understanding of the lesson concepts.

Inspiring Teacher Support

EL Support and suggested lesson alternatives are available throughout.

Question Prompts and Answers help support the conversation about STEM Connections.

Crosscutting Concepts Science Songs are available for Grade K–2 students.

Teacher suggestions on how to save time are included throughout.

Word Origin Study guides students' research through the word origins of the lesson vocabulary to better understand that parts of the words can give clues about the whole meaning.

Literacy and Math Connections are embedded throughout every lesson.

EL Support provides suggested activities for Emerging, Bridging, and Expanding student groups.

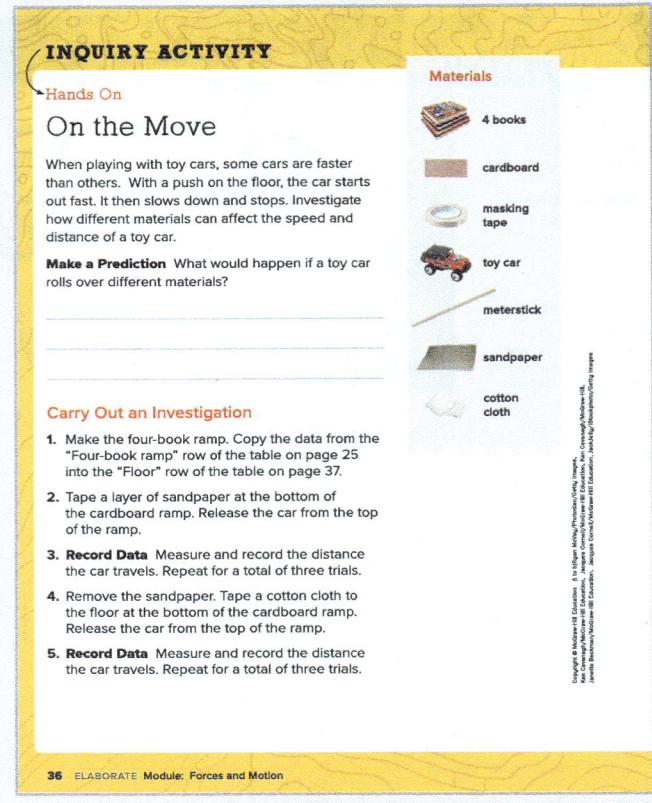

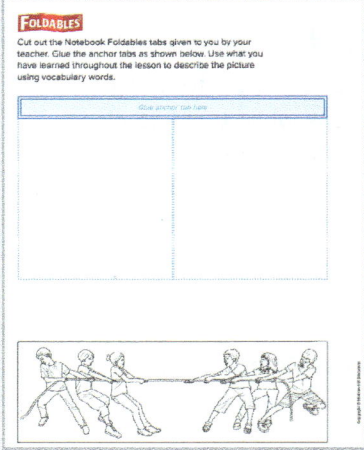

FOLDABLES

Use **Dinah Zike's Study Guide and Notebook Foldables**® as a tool to organize important lesson information and **Visual Kinesthetic Vocabulary**® to construct meaning and master lesson vocabulary.

49 Module and Lesson Walk Through

EXPLAIN ELABORATE EVALUATE MODULE PROJECT PLANNING MODULE PROJECT COMPLETION AND MODULE WRAP-UP

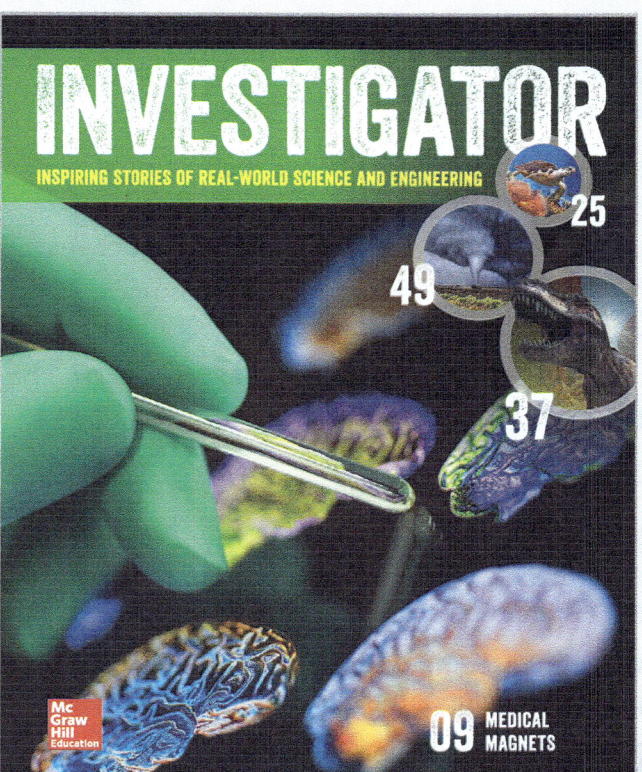

STEM Connections

Introduce students to real-world STEM professions that they may have one day. Students will learn about the career and then apply what they have learned to a related assignment.

INVESTIGATOR Articles

Students will engage with informational text and real-world science and engineering stories that are available in approaching level and on level.

Elaborate 50

Evaluate

MODULE OPENER — STEM MODULE PROJECT LAUNCH — LESSON LAUNCH SCIENCE PROBE — ENGAGE — EXPLORE

Evaluate
AT-A-GLANCE

In the Evaluate phase of the instructional model, you are able to gauge student progress toward achieving lesson objectives. This is a time to assess students' new understanding and abilities.

Inspiring Teacher Support

The **Environmental Connections** are identified in the Teacher's Edition.

Suggested activities are included to meet **ELD Standards**.

Differentiated Instruction suggestions are provided to support all learners.

Professional Learning Videos support your needs from start to finish.

Go online for interactive **Lesson Review** tools and resources.

The Online Assessment Center lets you assign students a pre-made **Lesson Check** that is based on the Disciplinary Core Ideas or customize your own practice assignments and assessments.

Students can practice important **21st Century Skills** with **Open Inquiry** activities.

Scoring Rubrics provide guidelines for the **Extend It** open inquiry activity.

EXPLAIN THE PHENOMENON

In the **Lesson Review**, students will demonstrate their learning by explaining the phenomenon, utilizing the SEPs and CCCs to showcase their Three-Dimensional Thinking skills, and extend their learning to real-world scenarios.

Lesson Checks and Interactive Practice

Assign students pre-made lesson checks and interactive practice tools that are purposefully designed to revisit the Disciplinary Core Ideas.

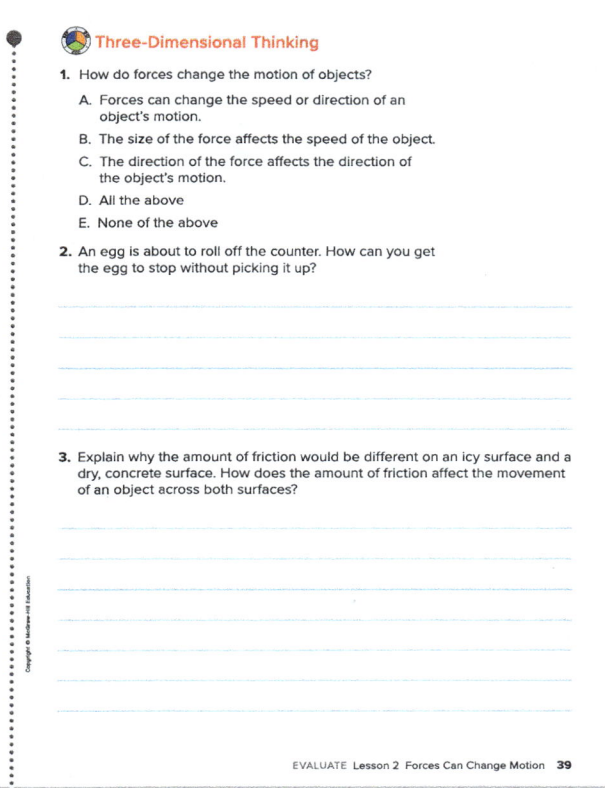

Three-Dimensional Thinking

Students will apply their three-dimensional learning to show their understanding.

See the Teacher's Edition for more three-dimensional thinking support and DOK levels.

ENVIRONMENTAL Connection

Environmental Connections help students to understand environmental impacts.

Extend It with Open Inquiry

Students engage in an open inquiry activity that focuses on 21st Century Skills.

STEM Module Project Planning

At the end of each lesson, students return to the STEM Module Project planning pages to apply what they have learned throughout the lesson to the STEM Module Project they complete at the end of the module.

Evaluate 52

STEM Module Project Planning

MODULE OPENER — STEM MODULE PROJECT LAUNCH — LESSON LAUNCH SCIENCE PROBE — ENGAGE — EXPLORE

STEM Module Project Planning
AT-A-GLANCE

Students in Grades 2–5 will use the **Project Planning Pages** at the end of each lesson to see how their learning can be applied to the **STEM Module Project** they'll complete at the end of the module. Students will define the problem they're trying to solve and complete research to deepen their understanding. They will think about the related STEM career that was introduced and discuss what real scientists or engineers do to answer science questions and prepare to solve a problem. After collecting the necessary information, they will sketch models and select the best one to build.

Inspiring Teacher Support

Project Parameters are clearly outlined and include student pages that should be revisited to help students with project planning.

Scripted facilitation questions are provided to guide student planning discussions.

Online **STEM Module Project** Teacher support pages that provide constraints, drawings, and additional support to teachers.

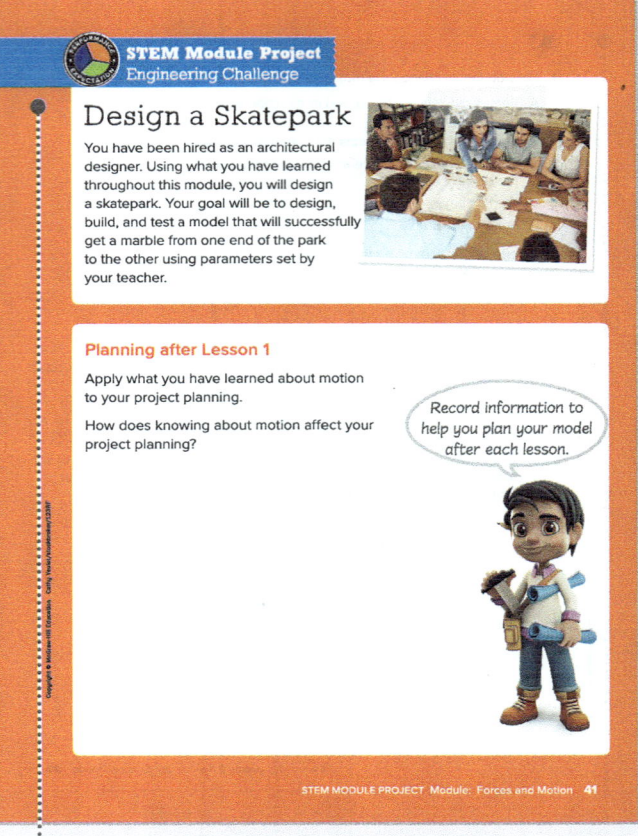

Review the STEM Module Project Parameters

After the first lesson in the module, students will revisit the purpose of the **STEM Module Project** and review how what they're learning will help with project planning.

For students in Grades K–1, detailed steps are provided to support their developmental needs.

EXPLAIN   ELABORATE   EVALUATE   **MODULE PROJECT PLANNING**   MODULE PROJECT COMPLETION AND MODULE WRAP-UP

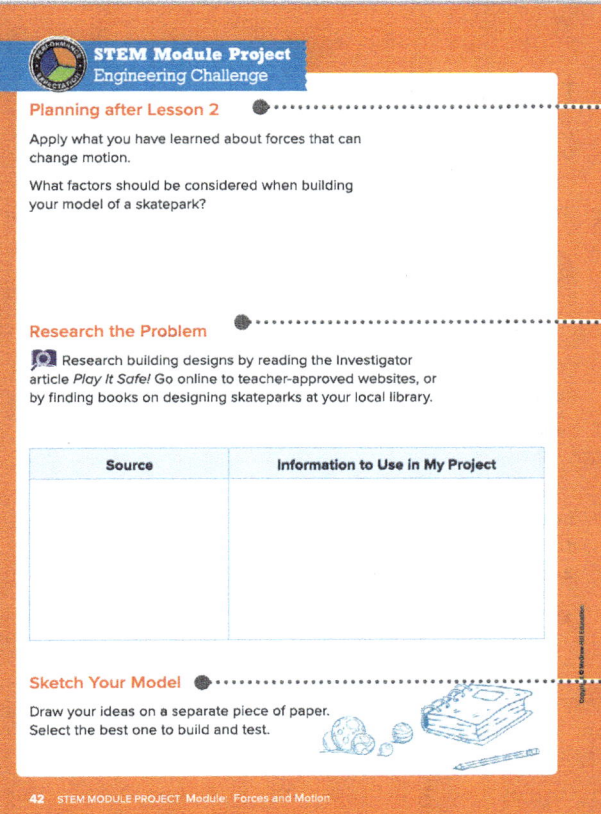

Lesson Planning Review

As they begin to complete their projects, students will revisit their planning notes from the close of each lesson.

Define the Problem and Complete Research

As part of the planning process, students will research possible materials they could use in their project.

Sketch Your Model

Before deciding on a final model to build, students are encouraged to sketch ideas on a separate piece of paper.

STEM Connection

During the project planning, students will review the related STEM career that was introduced at the beginning of the **STEM Module Project** and discuss what the professional's role would be at this point in the planning.

Module Project Rubric

Teacher and student rubrics allow students to decide on the criteria and constraints to assess their **STEM Module Project**.

STEM Module Project Planning   **54**

STEM Module Project Completion & Module Wrap-Up

MODULE OPENER — STEM MODULE PROJECT LAUNCH — LESSON LAUNCH SCIENCE PROBE — ENGAGE — EXPLORE

STEM Module Project Completion and Module Wrap-Up
AT-A-GLANCE

As the module comes to a close, students will complete a final culminating STEM Module Project to demonstrate their understanding of the Performance Expectations in the module. Through the completion of the project, students apply the three dimensions of learning to solve a problem related to the module phenomenon.

Inspiring Teacher Support

Background information and **STEM Connection** support is provided to connect the STEM Module Project to real-world STEM projects.

Scripted questions are provided to support group discussion facilitation.

Communicate Your Results support helps you guide students to the best way to communicate their project results.

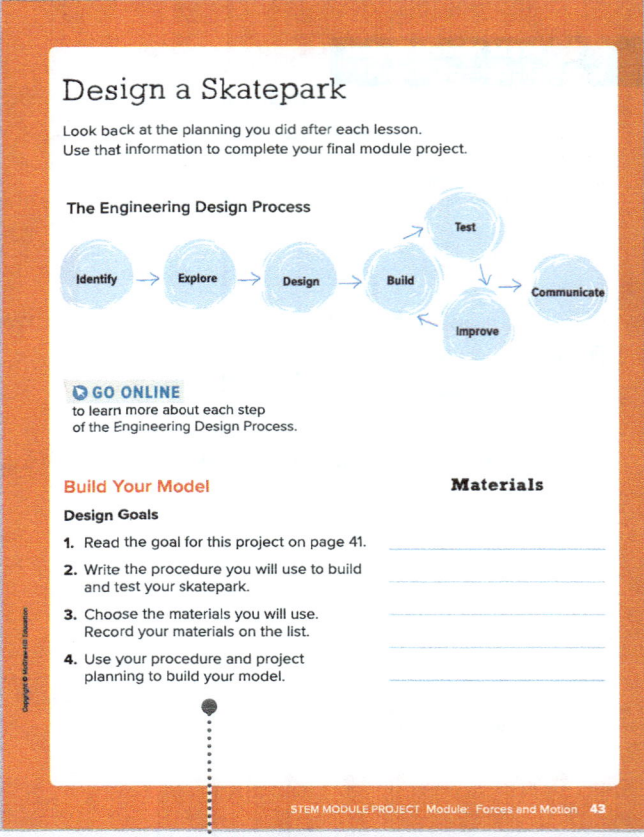

1. Build Your Model

- Review the Design Goals.
- Prepare a list of materials needed to build the model.
- List the procedure used to design the model.
- Build the model they designed.
- Test, record observations, and make improvements.

55 Module and Lesson Walk Through

Inspire Science

EXPLAIN — ELABORATE — EVALUATE — MODULE PROJECT PLANNING — **MODULE PROJECT COMPLETION AND MODULE WRAP-UP**

2. Test Your Model and Communicate Your Results

Students should refer to their rubric at the beginning of the project to make sure their model fits the criteria.

Online eAssessment Center
 GO ONLINE

Assign a premade Module test based on the Disciplinary Core Ideas or customize your own test.

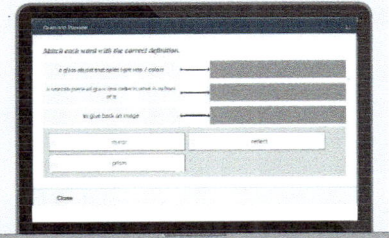

Module Wrap-Up

Students revisit the module phenomenon and try to answer the phenomenon question using evidence from what they have learned throughout the module and the STEM Module Project.

STEM Module Project Completion and Module Wrap-Up **56**

Digital Experience

Inspire Science
Digital Experience

Immerse yourself in the *Inspire Science* digital experience. This section will provide an overview of the following:

* Course Dashboards
* Module and Lesson Landing Pages
* Digital Resource Types and Learning Impact

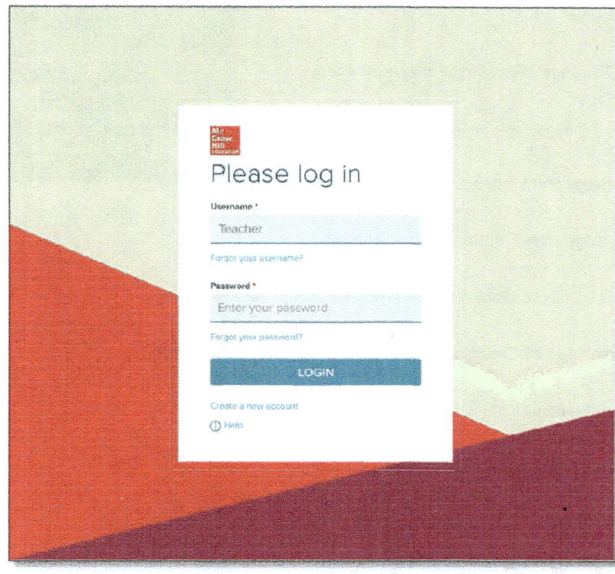

Get Started by Logging In:

1. Go to https://my.mheducation.com from an Internet browser.
2. Enter your username and password and click "Log In."

Upon login, you will find helpful videos to support your digital review.

 **Need login credentials?**
Go to my.mheducation.com and select "Create Teacher Account."

The digital designs and navigation shown in this guide may vary as we continue to enhance the digital experience.

Digital Experience

Welcome to the *Inspire Science* digital experience!

Use this section of your Program Guide to easily find the digital resources that make *Inspire Science* engaging and fun for students.

Browse Your Course

Upon login, you will see a colorful banner for your course showing the images from your book covers. Select "Browse Your Course" or click anywhere in this banner to begin accessing your course resources.

Choose a Module and Lesson

After launching your course, you will land on the table of contents page with expandable folders for all modules and lessons in the course, as well as folders with documents to support understanding of the entire program, such as this Program Guide. Select a module, or a lesson within a module, to access the module and lesson landing pages.

> Select a module or lesson to access the module and lesson landing pages, where you will find resources such as planning tools, professional learning resources, and student resources aligned to the print Student Editions.

- ∨ Browse Your Course
  - Program Overview: Welcome to Inspire Science
  - Program Resources: Course Materials
  - › Module: Energy and Motion (Unit 1: Forces and Energy)
  - › Module: Energy Transfer (Unit 2: Using Energy)
  - › Module: Natural Resources in the Environment (Unit 2: Using Energy)
  - › Module: Earth and Its Changing Features (Unit 3: Our Dynamic Earth)
  - ∨ Module: Earthquakes (Unit 3: Our Dynamic Earth)
    - Lesson 1: Map Earthquakes
    - Lesson 2: Model Earthquake Movement
    - Lesson 3: Reduce Earthquake Damage
  - › Module: Structures and Functions of Living Things (Unit 4: Information Processing and Living Things)
  - › Module: Information Processing and Transfer (Unit 4: Information Processing and Living Things)
  - › Additional Resources: Beyond the Classroom
  - Program Resources: Glossary

Note: Digital design and navigation may vary.

Inspire Science

Access Module Interactive Resources

Module Landing Pages

From the module landing pages, you can access module resources for teachers and students, organized by key module-level activities. Module-level resource folders for each module include:

- Module Planning Resources (including Professional Learning Resources)
- Module Opener
- STEM Module Project
- Module Wrap-Up
- Module Assessment
- Module Library (including leveled readers and additional STEM career connections)

Easily navigate to other module and lesson landing pages by using the module and lesson drop-down menus.

To collapse or open sections, click on ⌄

Note: Digital design and navigation may vary.

Digital Experience

Digital Experience

Access Your Resources

You will notice within the module and lesson landing page folders that many digital resources are further organized by two categories:

1 Interactive Presentation

These resources provide access to the digital content that aligns with the resources featured in the print Student Edition. By default, these resources will display on the student page and in the teacher presentation. Resources in the Interactive Presentation section of the module and lesson landing page is optimized for digital projection and student 1:1 device use.

2 Additional Resources

These resources provide access to supplemental content, optional content, and assessments. Resources in this section are typically hidden from students until teachers are ready to add them to student pages or assign them.

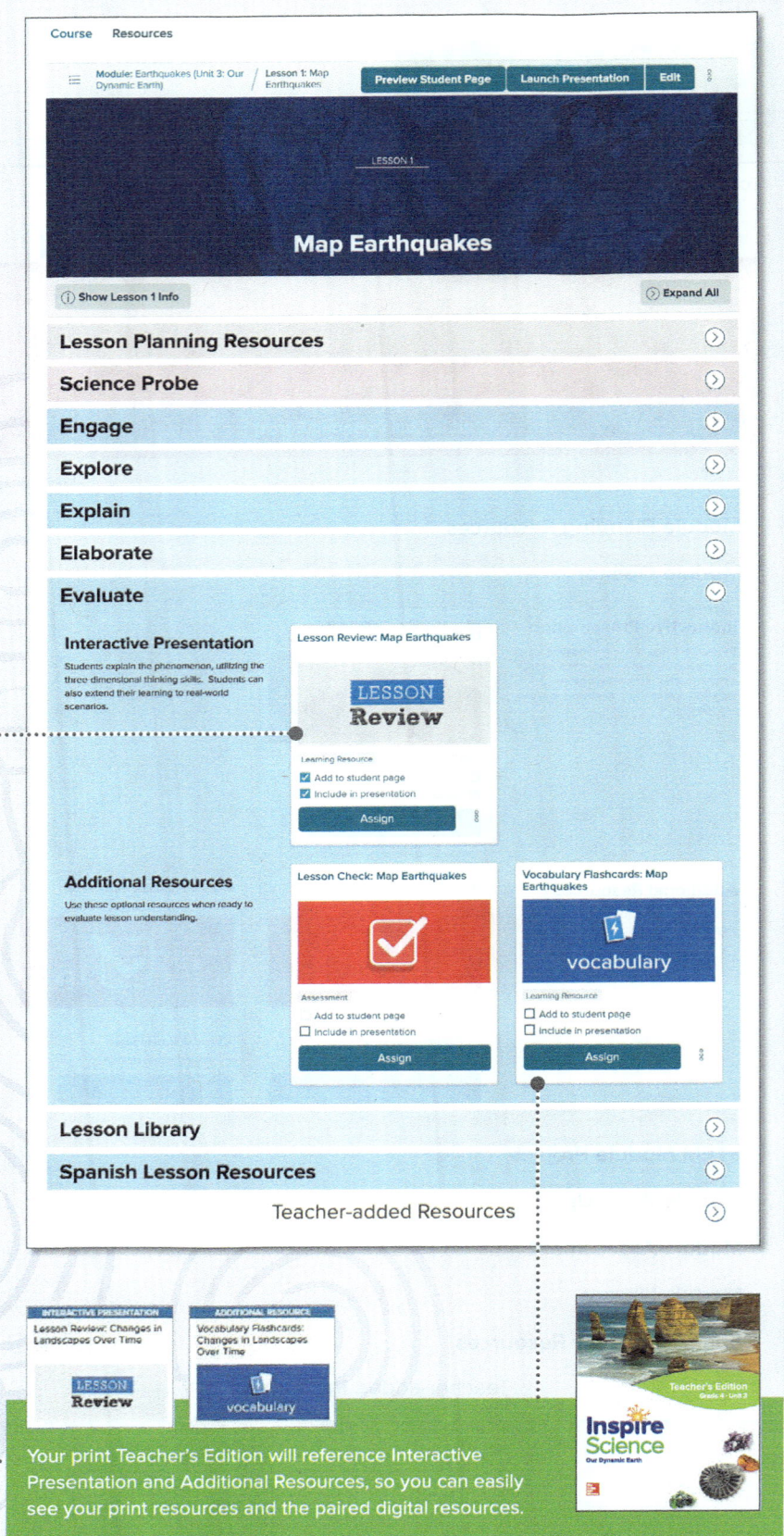

Your print Teacher's Edition will reference Interactive Presentation and Additional Resources, so you can easily see your print resources and the paired digital resources.

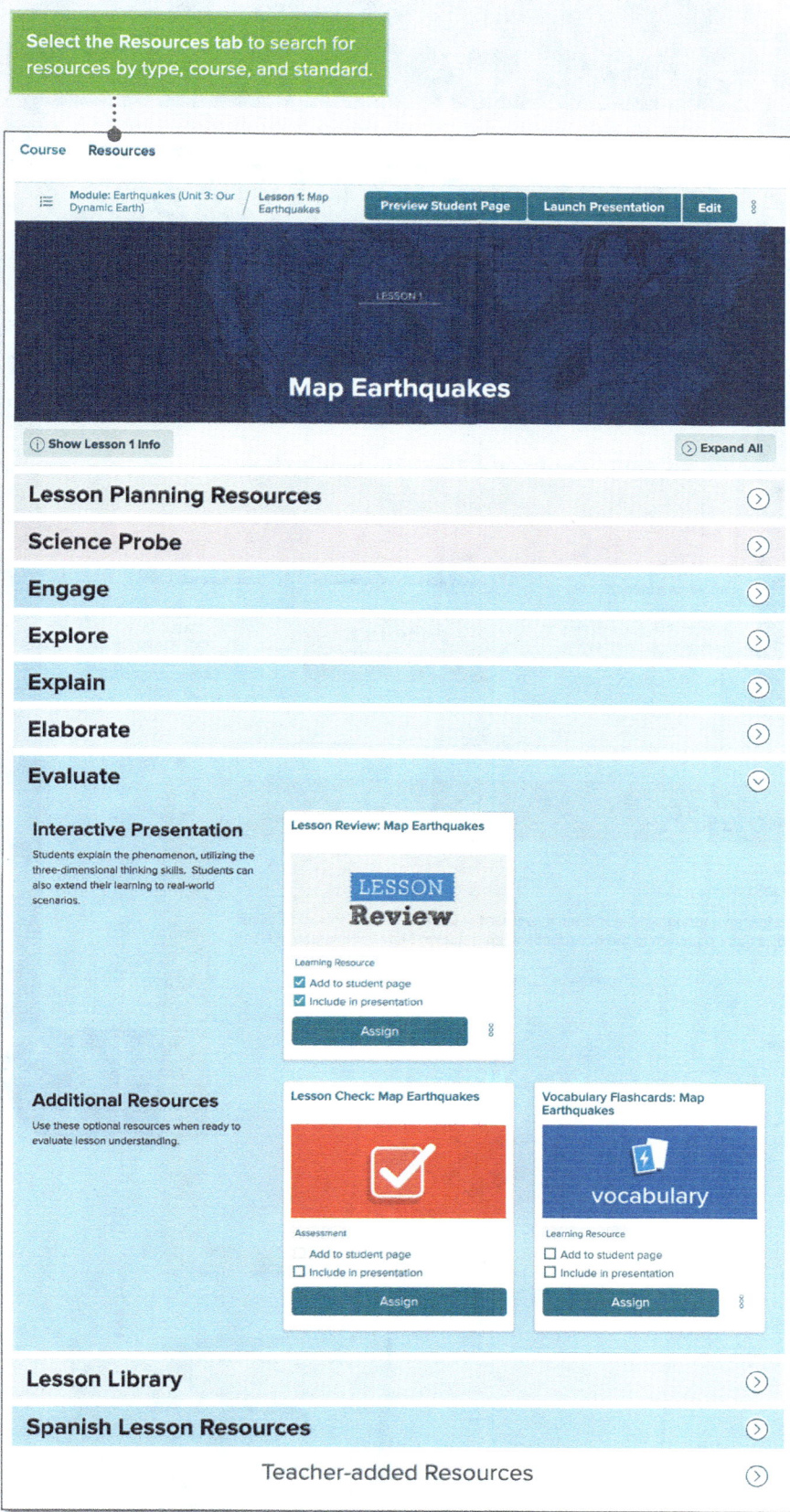

Select the Resources tab to search for resources by type, course, and standard.

Access Lesson Interactive Resources

Lesson Landing Pages

From the lesson landing pages, you can access lesson resources for teachers and students, which are organized by the 5E instructional model. Lesson resource folders for each lesson include:

- Lesson Planning Resources
- Science Probe (Formative Assessment)
- Engage
- Explore
- Explain
- Elaborate
- Evaluate
- Lesson Library

Accessing Course Resources

Digital Experience

Viewing Digital Resources

Inspire Science offers a variety of rich media and interactive content with the flexibility to customize lessons to fit your needs.

Follow these tips for viewing resources:

1. Select
From a landing page, select any resource to launch and review it.

2. View
While reviewing a resource, use the red arrows to navigate through the screens of each resource.

3. Close
Once you are finished reviewing, close out by selecting "X" to get back to the landing page.

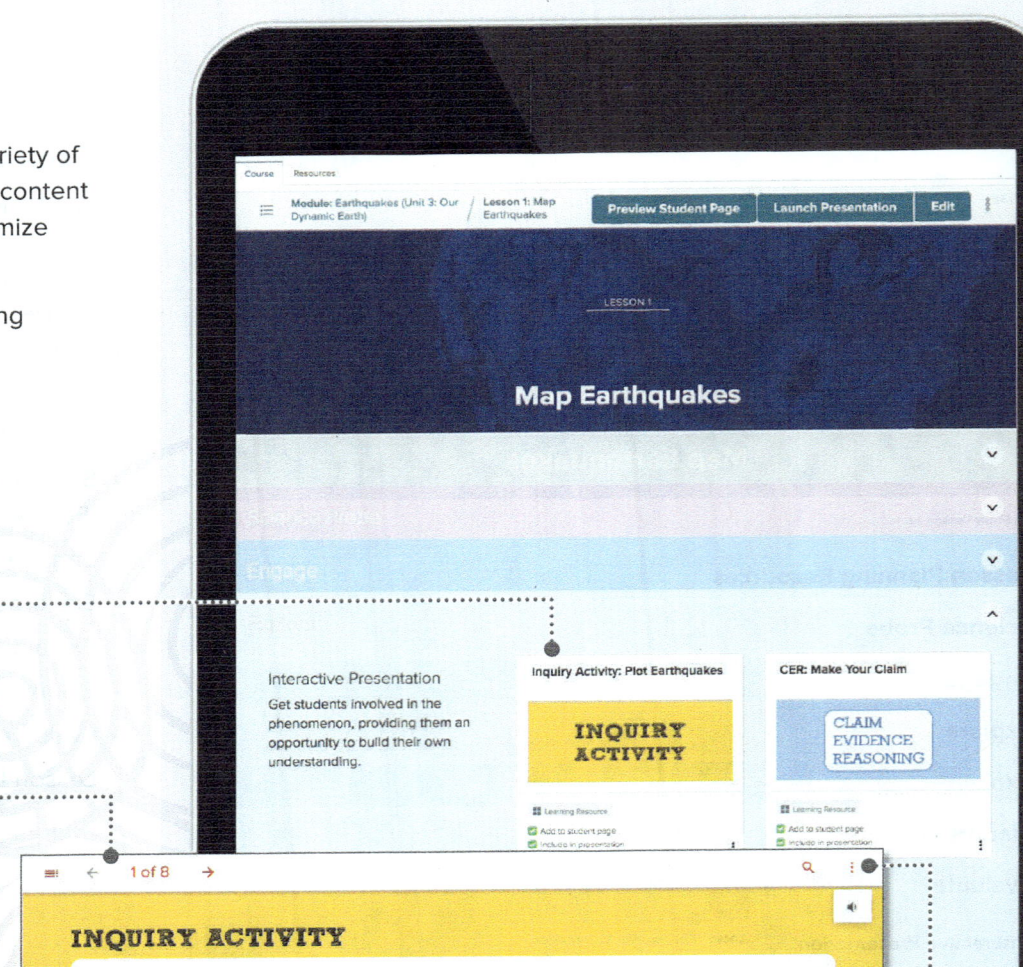

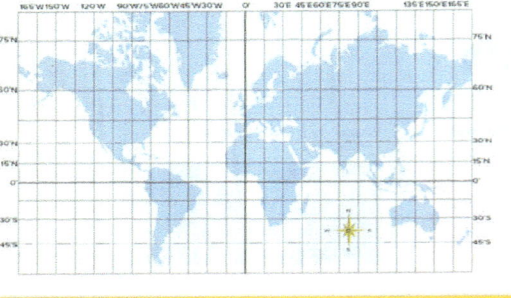

To reset an activity within a resource (clear any content entered), use the three vertical dots and select "Reset Activities."

62 Digital Experience

Note: Digital design and navigation may vary.

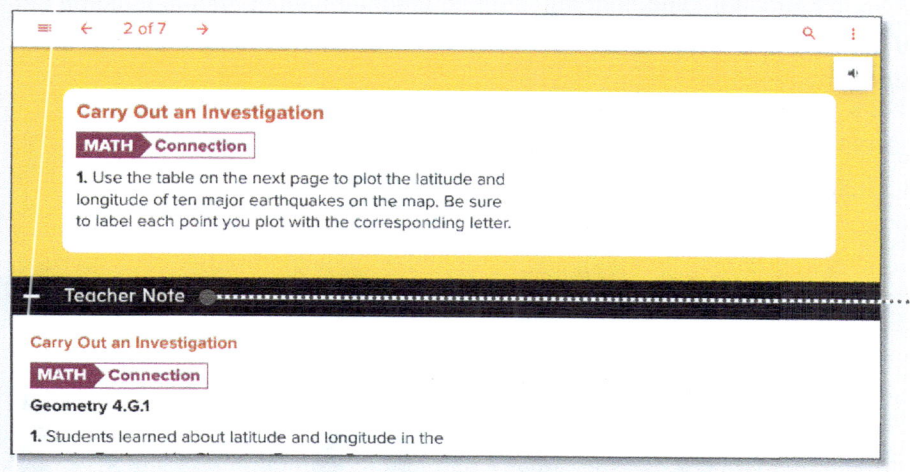

Teacher Note
From the Online Teacher Center login, teacher support can be seen at point of use by expanding (select +) the Teacher Note section.

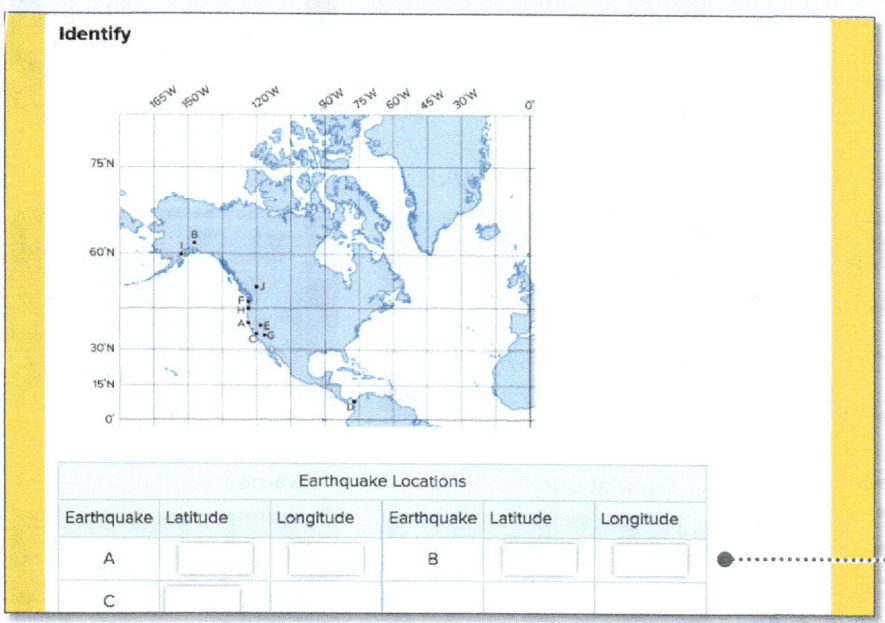

Table Entry
Students can enter data into tables at point of use for review.

Audio Support
Select the speaker icon to hear the on-screen text read aloud.

Answers
From the Online Teacher Center login, answers can be seen at point of use by expanding (select +) the Answer section.

Digital Experience

Types of Interactive Resources

In the *Inspire Science* digital experience, students will interact with a wide variety of digital content types that will make learning science engaging and fun.

Engaging Online Resources

The following list is a few of many offerings for *Inspire Science*:

- Engaging interactive content
- Video demos of hands-on activities
- Science content videos
- Text read aloud and highlighting features
- Dynamic search tools
- Impact News

Drawing Tool

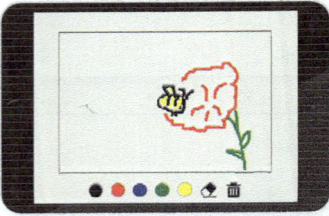

Drag and Drop

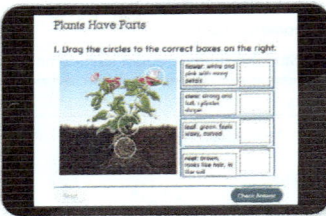

Phenomena Videos

Science Content Videos

Pop Tips

Layer Reveal

Simulations

Games

Impact News

Choose Your Path

Interactive Text

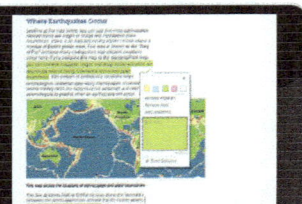

Beyond the Classroom (2-5)

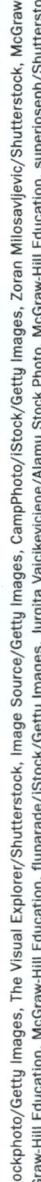

(l to r, t to b)iStockphoto/Getty Images, The Visual Explorer/Shutterstock, Image Source/Getty Images, CampPhoto/iStock/Getty Images, Zoran Milosavljevic/Shutterstock, McGraw-Hill Education, McGraw-Hill Education, flyparade/iStock/Getty Images, Jurgita Vaicikeviciene/Alamy Stock Photo, McGraw-Hill Education, superjoseph/Shutterstock

Inspire Science

Drawing Tool

The **Drawing Tool** allows students to illustrate responses and annotate images for their assignments. Students can also use the drawing tool to analyze and graph data.

Impact News

Impact News is a current events news site that provides two news stories, in both English and Spanish, that are published monthly and curated specifically for Inspire Science. Each article is written at three readability levels to provide differentiation support for all learners.

Digital Interactive Tour   65

Digital Experience

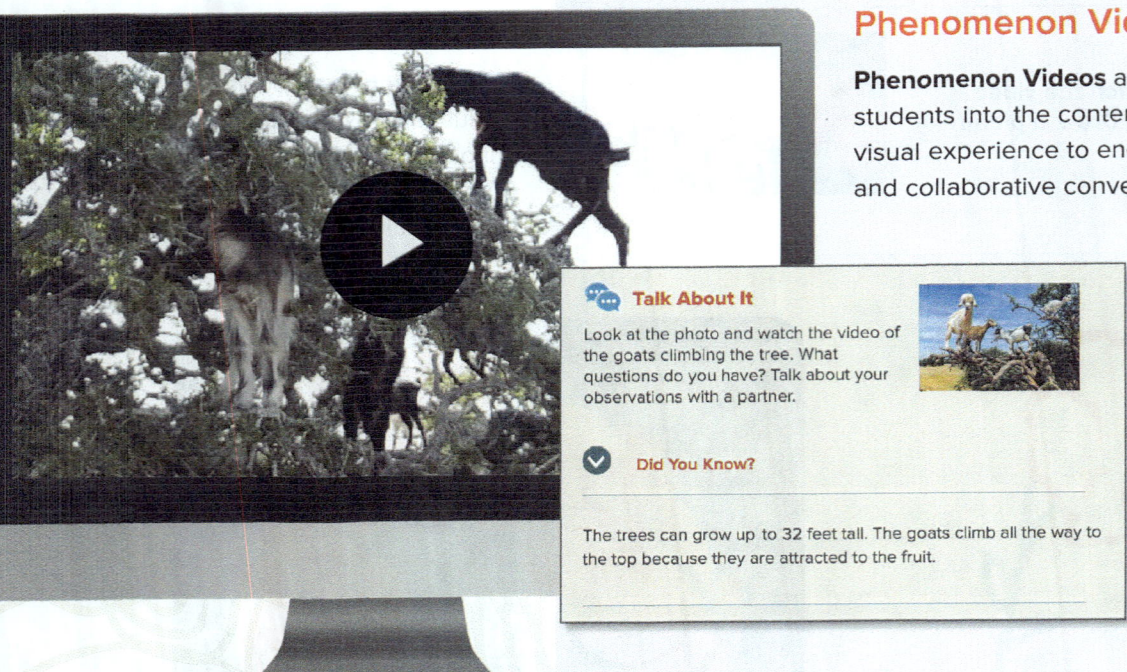

Phenomenon Videos

Phenomenon Videos are used to draw students into the content and provide a visual experience to encourage thinking and collaborative conversations.

Science Content Videos

Bring interesting phenomena to life and enable students to feel like they are a part of the experience with inspiring **Science Content Videos**.

Inspire Science

STEM Videos

Real-world STEM Connection videos and STEM Career Kid videos (K-1) introduce a variety of interesting science and engineering professions.

Professional Learning Videos

Inspire Science comes with library of relevant, self-paced, **Professional Learning Videos** and modules to support you from implementation through ongoing instructional progression.

Digital Interactive Tour

Digital Experience

Flash Cards

Flash Cards are used to present information with interactive text. Vocabulary Flash Cards include the vocabulary term on one side and the definition on the back. Activity flashcards can be used to present images and text describing before and after events.

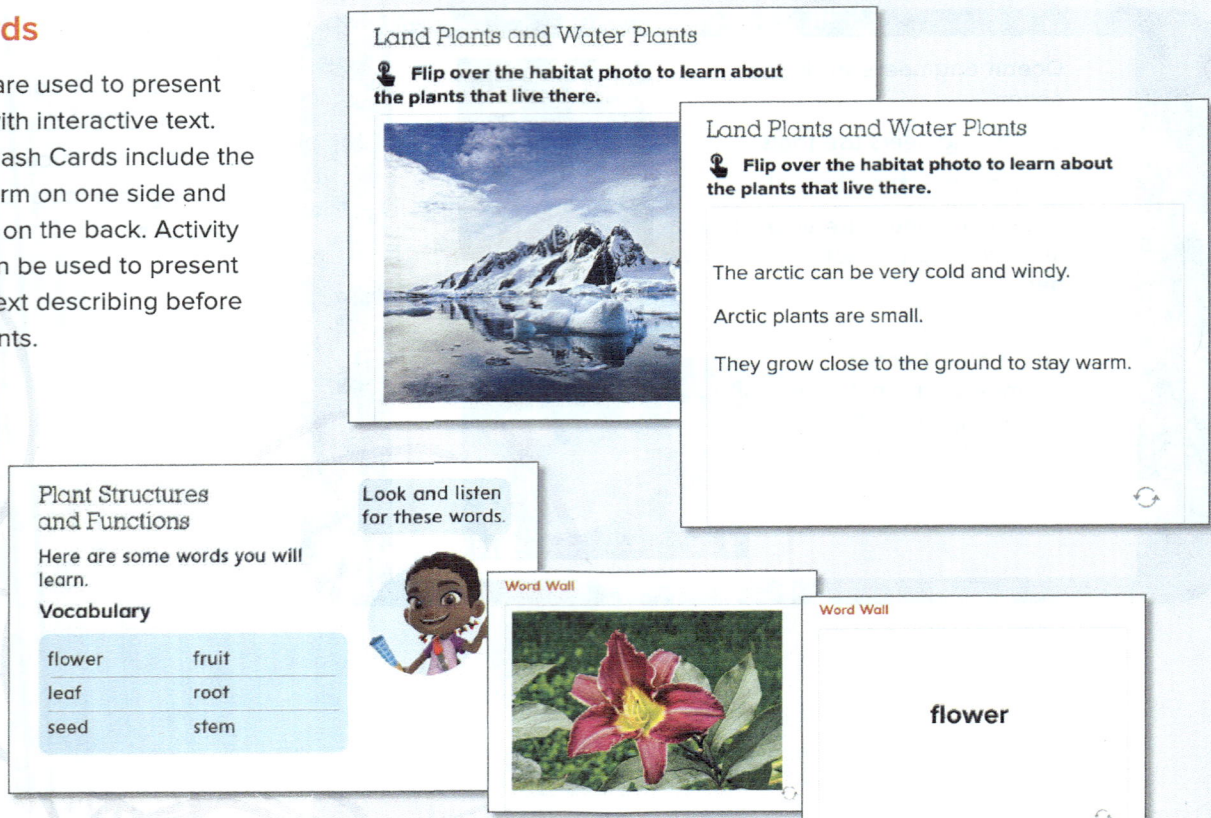

Pop Tips

Pop Tips allows students to interact with images and connect to related information in order to support understanding of core content.

Inspire Science

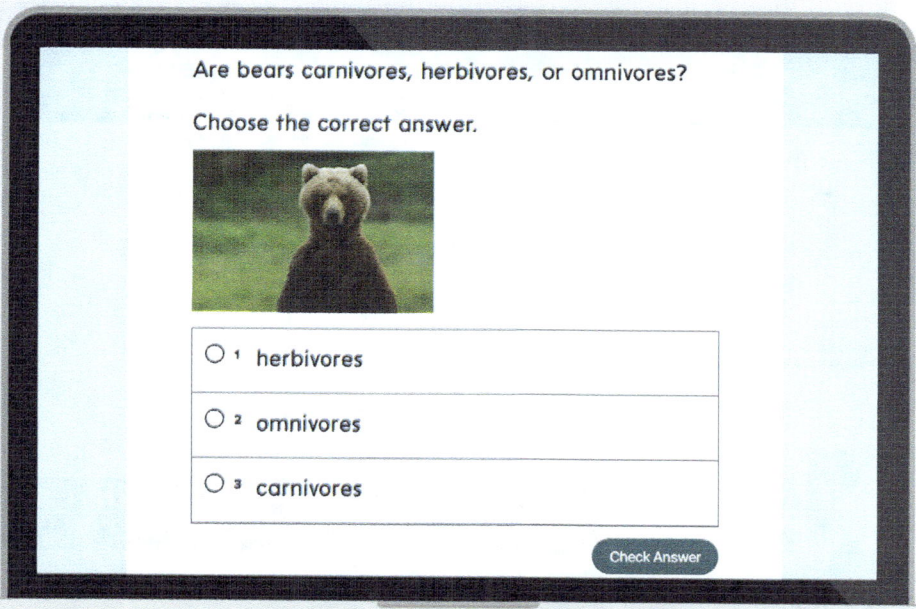

Multiple Choice

The **Multiple Choice** interactive is ideal for classifying content, making a claim, identifying key terms, and conducting formative assessment.

Layer Reveal

The **Layer Reveal** interactive enables students to easily visualize cause-and-effect scenarios and focus on specific areas of an image, one focused section at a time.

Digital Interactive Tour

Digital Experience

Simulations

Simulations are used to provide students an experience when the activity isn't easily replicated in the classroom with a hands-on inquiry activity.

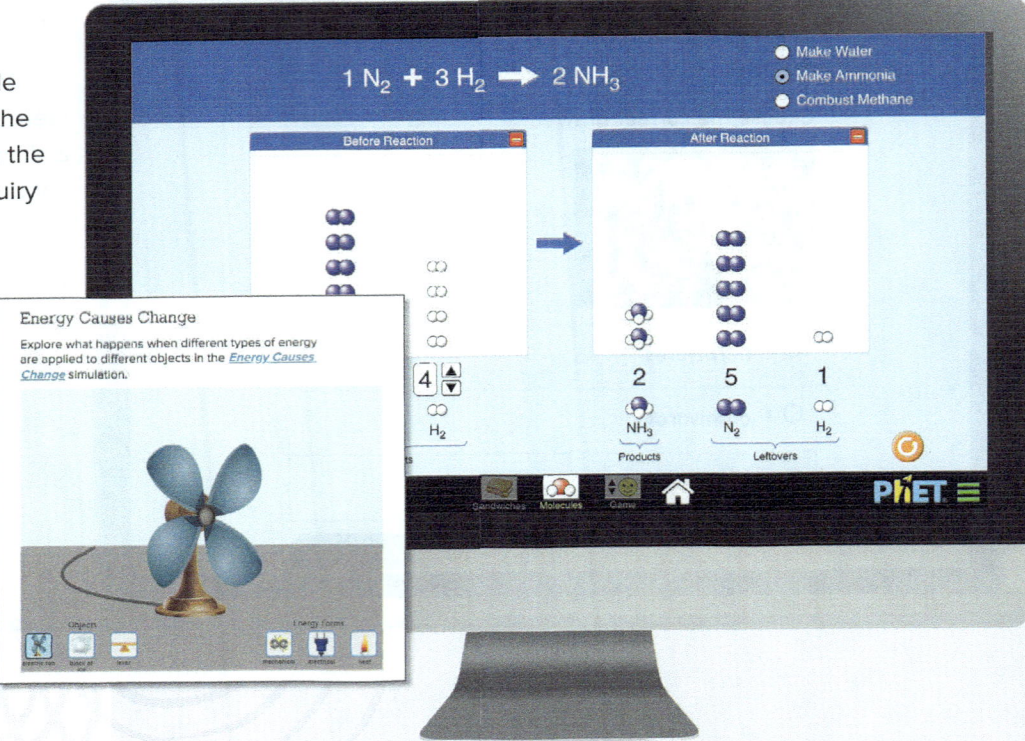

Games

Digital learning games reinforce deeper conceptual science understanding by immersing students in experiential learning.

Inspire Science

Choose Your Own Path

The **Choose Your Own Path** interactive enables students to direct their own learning experience.

Slide Line Plus

The **Slide Line Plus** feature allows students to progress through a storyline of images or highlight focused areas of visuals to concentrate on one element of a schematic at a time.

Digital Interactive Tour 71

Digital Experience

Personal Tutors

Students have access to **Personal Tutors** when they need extra support learning new concepts.

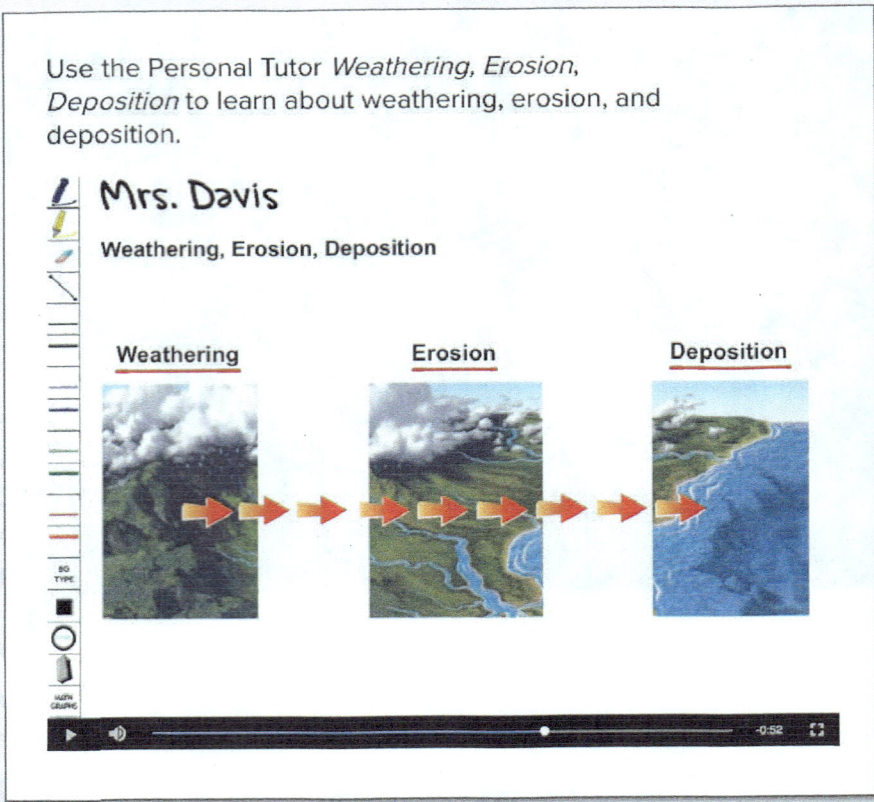

Click Change

The **Click Change** interactive is used to allow students to engage with images. Students might click through images to select the correct one in a vocabulary check or click through images in an activity to identify similarities and differences.

Inspire Science

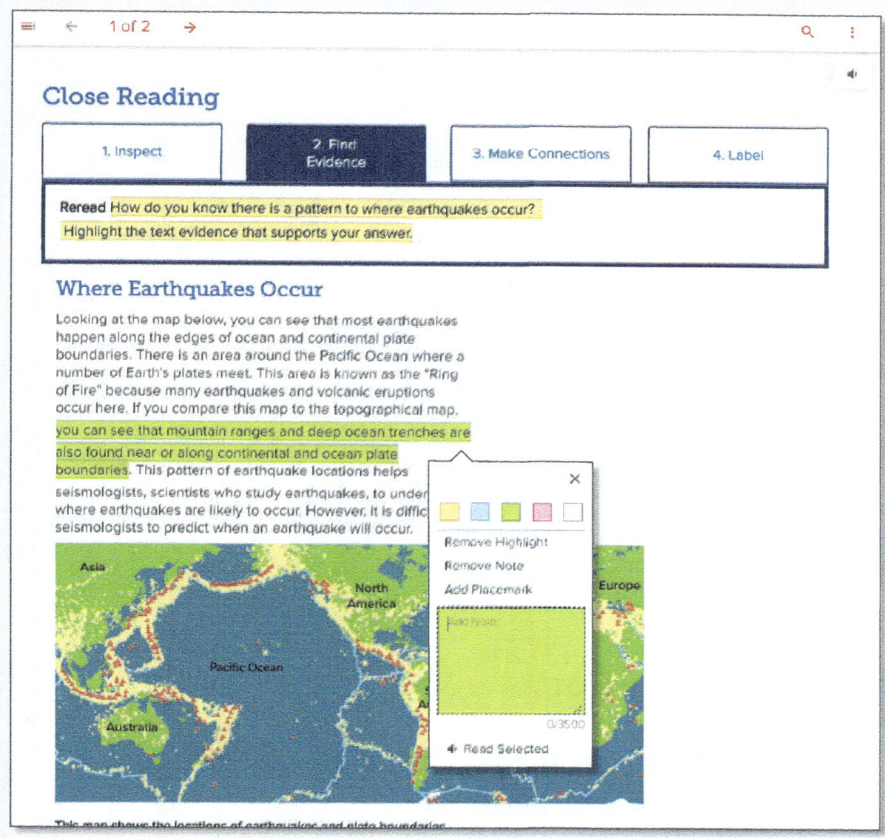

Interactive Text

Students become more engaged in close reading activities with **interactive text** features:

- Text highlighting
- Place marking capabilities
- Note-taking
- Text-to-speech reading

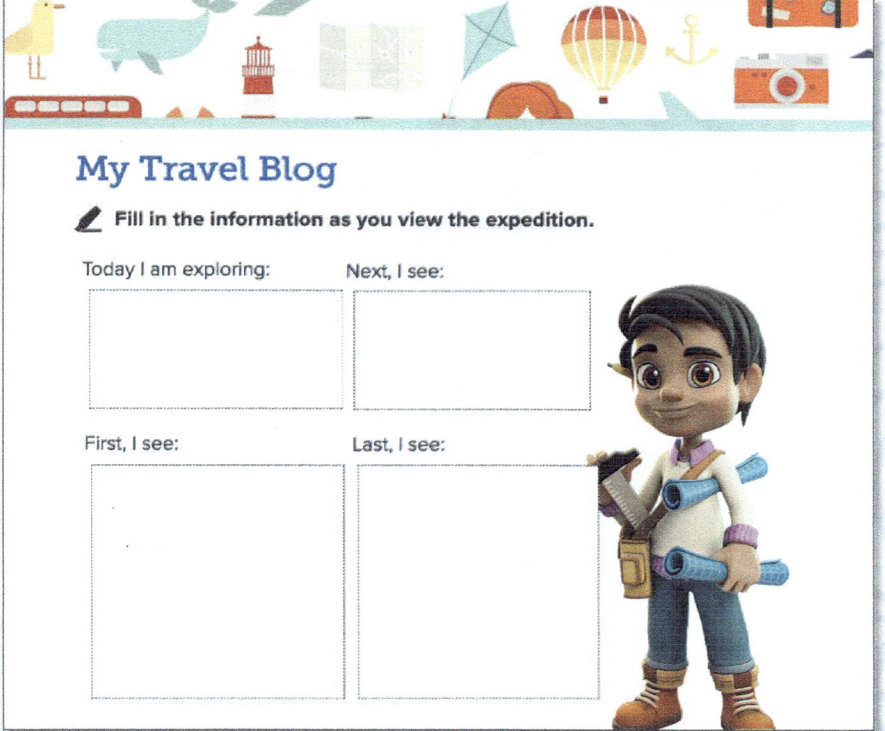

Beyond the Classroom

Beyond the Classroom is a virtual field trip experience. It provides students tools to help document their Google Expeditions® journey.

Digital Interactive Tour

Digital Experience

Type Entry

Students can record, edit, and save their assignment responses.

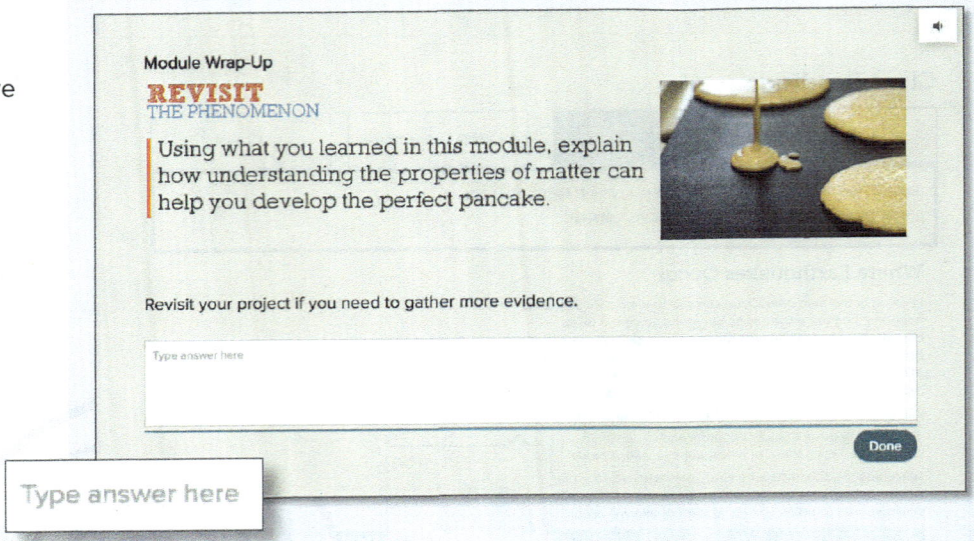

Drag and Drop

The **Drag and Drop** interactive is used to support students with sorting and classifying content such as vocabulary terms.

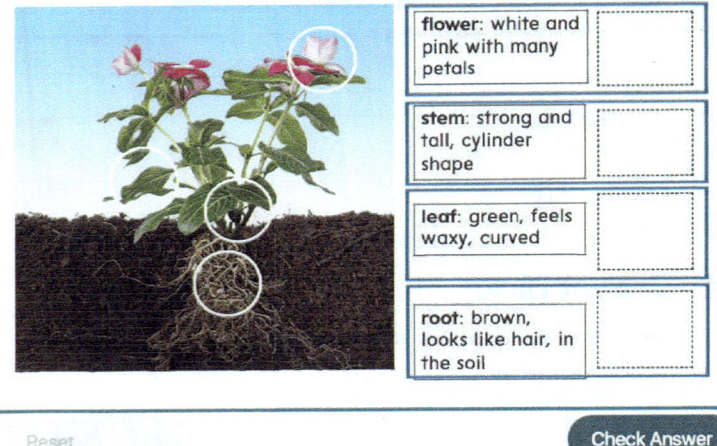

Inspire Science

Thank you for all you do to inspire your students to be curious, to investigate, and to innovate.

Let's Explore Our Phenomenal World!